Fertilizers and manures

Ken Simpson

Longman Scientific & Technical

Copublished in the United States with
John Wiley & Sons, Inc., New York

Longman Scientific & Technical
Longman Group Limited
Longman House, Burnt Mill,
Harlow, Essex CM20 2 JE, England
and Associated Companies throughout the world

*Copublished in the United States with John Wiley & Sons
Inc., 605 Third Avenue, New York, NY 10158*
© Longman Group Limited 1986

First published 1986
Reprinted 1991

British Library Cataloguing in Publication Data
Simpson, Ken, 1921–
  Fertilizers and manures.——(Longman handbooks in
  agriculture)
  1. Fertilizers   2. Manures   I. Title
  631.8   S633

ISBN 0-582-44747-X

Library of Congress Cataloging in Publication Data
Simpson, Ken, 1921– Fertilizers and manures.
  (Longman handbooks in agriculture)
  Bibliography: p.
  Includes index.
  1. Fertilizers. 2. Manures. I. Title. II. Series.
S633.S624 1986   631.8'1   85–15950

  ISBN 0–470–20693–4 (USA only)

Produced by Longman Group (FE) Limited
Printed in Hong Kong

Contents

Preface

The aim of this book is to encourage the most effective use of fertilizers and manures. These materials, judiciously used, have been an essential factor in the great developments which have occurred in crop production during this century leading to the prodigious improvements in both yield and quality.

The early chapters give essential basic information on plant nutrition and the soil processes which affect the availability of nutrients to the plant. Later chapters are concerned with the application of this knowledge to the management of fertilizers and manures in the interests of efficient crop production, minimum wastage of nutrients and the control of pollution.

Lime is considered as a key fertilizer because of its critical rôle in the control of pH and nutrient availability. Adjustments to traditional liming practices are suggested to meet the needs of intensive crop production.

Manures, and especially slurry, are commonly regarded by the farmer simply as posing disposal problems. There are major advantages in the effective fermentation and storage of manures and their incorporation into the soil. The simple measures required to ensure efficient use of these manures have been set out. The value of returning all surplus crop residues to the soil, especially straw and green manure crops, is also stressed.

The last chapters of the book deal with all aspects of fertilizers and their use, especially the many factors which affect the amount and types of fertilizer required by the crops. The term 'simple' fertilizer has been used throughout to describe fertilizers designed to supply only one plant nutrient; the term 'compound' is applied to fertilizers supplying two or more nutrients. The properties of all the main fertilizer materials are set out simply to illustrate how they affect the choice of fertilizer for particular purposes.

In addition to the ubiquitous NPK, attention is drawn to the critical importance of sulphur and magnesium as major plant nutrients and the hazards of neglecting these elements when devising fertilizer policies. The functions of trace elements and problems of prevention and control of excess and deficiency are also considered.

The nature of yield responses of crops to increasing rates of fertilizer is discussed in detail and especially the consequences of applying excessive amounts of fertilizer. Abuses such as excessive applications or unsuitable methods of application often have adverse effects on both crop yield and quality, and on soil properties, and also increase stream pollution. These problems are stressed and a chapter is devoted to the commonly neglected effects of fertilizers on the various aspects of crop quality.

No attempt has been made in this book to give comprehensive fertilizer recommendations for crops. Such recommendations are ably devised by government-sponsored advisory services and by reputable fertilizer firms. They are subject to frequent changes resulting from new knowledge and especially from the introduction of new crop varieties. Also they need to be interpreted and adjusted in the light of detailed local knowledge. To give specific recommendations in a book of this type could be very misleading. I have, therefore, concentrated upon the significance of the factors which need to be taken into account when deciding on fertilizer policy and have given particular emphasis to many simple ways of economizing on fertilizers without detriment to crop yields or quality.

The final chapter looks forward to developments in fertilizer technology and use and calls into question the goal of maximum yields, on grounds of economic uncertainties, pollution risks, effects on crop quality and the need to conserve limited reserves of raw materials used in fertilizer production.

As in the companion book *Soil* published in 1983, 'jargon' terms have been avoided wherever possible. A separate index of essential terms has been included to guide the reader to pages on which key definitions or descriptions are given.

Metric units of length, area and weight have been used throughout. No attempt has been made to present equivalent imperial measurements. For those who wish to use yards, acres and pounds a simple conversion table is given (p. 243).

The original intention was to express all plant nutrients in terms of the element. This has been thwarted in the case of phosphorus (P), potassium (K) and magnesium (Mg) by the dominant and consistent quotations in the British and EEC fertilizer industries and regulations of phosphorus pentoxide (P_2O_5), potassium oxide (K_2O) and magnesium oxide (MgO) instead of the elements. I have, therefore, with great regret, decided to express quantities of these three elements in terms of their oxides.

Apart from my own knowledge of fertilizers and manures, gained over a period of forty years research, development, teaching and advisory work, I have consulted a wide range of publications. Many of the figures and tables are derived from my own research or from data collected in the course of advisory work. Others have been constructed using information brought together from several sources or reproduced directly or redrawn from the data of a single published work. The sources in each case are gratefully acknowledged.

I wish to record my warmest thanks to Dr Phil Crooks and Mr Ron B. Speirs. Both have been for many years my close collaborators in the research, development and advisory work, the results of which are an essential part of this book. During its preparation they have supplied much valuable information and freely discussed with me the problems which have risen. I am also deeply indebted to Mr Donald McKelvie, Assistant Director of Extension Services, Edinburgh School of Agriculture, for his detailed reading and criticism of the script from the agriculturalist's viewpoint. Many other members of staff of the School have been plagued by me. My thanks are due in particular to Dr Keith Chaney, Dr Carol M. Duffus, Dr Alun Edwards, Dr David Lockhart and Mr Kevin Volans as well as Mr Duncan Hunter, farmer and former student, for many valuable discussions on the practical aspects of fertilizer use.

Some of the information contained in the many invaluable advisory bulletins and leaflets of the Agricultural Development and Advisory Services of England and Wales and those of the East of Scotland College of Agriculture has been used directly or indirectly during the preparation of this book. It is impossible to overestimate the value of these publications and I strongly recommend them to my readers.

Several producers, manufacturers and sellers of fertilizers, lime, manures and distribution machinery have willingly supplied information and photographs. In particular, I wish to thank for their

co-operation: Messrs A. C. Bamlett Ltd, Thirsk, North Yorkshire;
J. W. Chafer Ltd, Doncaster, South Yorkshire; Lely Import Ltd,
St Neots, Huntingdon, Cambs; Norsk Hydro Fertilizers, Ipswich,
Suffolk; John Parsons Marketing Ltd, King's Lynn, Norfolk;
Saltney Engineering Ltd, Bishops Stortford, Herts; UKF
Fertilizers Ltd, Ince, Chester; Vicon Ltd, Ipswich, Suffolk; and
John Wilder (Engineering) Ltd, Wallingford, Oxon.

Ken Simpson
February 1985

Longman Handbooks in Agriculture

Series editor:

C. T. Whittemore

Books published

J. D. Leaver: *Milk production – science and practice*
A. W. Speedy: *Sheep production – science into practice*
K. Simpson: *Soil*
E. Farr and W. C. Henderson: *Land Drainage*
J. Hodgson: *Grazing Management*
C. T. Whittemore : *Elements of Pig Science*

Chapter 1 The need for fertilizers and manures

Fertilizers and manures are used in agriculture to supplement the nutrients which the plant can obtain from the soil alone. The result is usually an increase in yield, sometimes spectacular. Another aim of the use of fertilizers and manures, not always successful, is to improve the quality of the crop as food for human beings and other animals.

For efficient growth the plant needs a range of essential elements in addition to the carbon, hydrogen and oxygen which compose most of it. Some, such as nitrogen and potassium, are needed in large quantities and are, therefore, described as major elements. Others, including among them copper and molybdenum, are needed in smaller amounts and are known as trace elements. The term 'minor' is sometimes used instead of 'trace' element but it is not a good term as it infers a minor rôle in plant nutrition which is untrue. Deficiency of a trace element can kill a plant just as surely as a lack of nitrogen.

Manures   The term 'manure' is used in this book to describe bulky organic materials, mainly plant residues and animal excreta, which are returned to the soil either directly or after some sort of processing. In farmyard manures and slurries this is the passage of plant material through the animal and subsequent fermentation.

Manures commonly contain much water – slurries as much as 95 per cent and farmyard manures about 75 per cent – and the concentration of plant nutrients in them is low. As a result large quantities (25 t/ha or more) are needed to supply an appreciable part of the nutrient requirements of the plant.

By their nature, manures have two functions. They supply some organic matter to the soil, much of which is lost to the atmosphere after conversion to carbon dioxide, but some of which is changed to humus – a black or dark brown organic substance which persists in the soil and improves its physical properties. Manures also supply a wide spectrum of plant nutrients derived from the crop residues of which they are composed. Well conserved farmyard manures will contain the full range of essential nutrients although not necessarily in the proportions needed for crops.

Manures most widely used at present are animal products, traditional farmyard manures made with straw, or the more recent slurries made by simply diluting animal excreta with the water used for cleansing the living quarters of stock.

The value of animal manures for improving crop production has been recognized for a very long time. Both Greek and Roman writers, in the late centuries BC and early AD, discussed in detail the relative merits of the excreta of horses, cows, goats, sheep, poultry, other birds and man for different soils and crops.

Animal manures have been widely used ever since those times, partly because of the need to dispose of them, but with an increasing appreciation of their short- and long-term value. There is striking evidence of the long-term value in the dark colour, soil structure and fertility of the 'in-by' fields which surround very old-established farmsteads.

It was not until the sixteenth century that the need to conserve animal manures for maximum efficiency was fully recognized. At that time Palissy gave a description of the precautions needed to preserve farmyard manure similar in all essential details to those given in this book. He fully appreciated the need for cover to avoid losses by leaching and volatilization, for storage 'in a shallow cavity like a small pond . . . paved with bricks or stone using a mortar of lime and sand'. He also prized the liquor which accumulated in the storage area and wrote that 'It should be conveyed to the field in water-tight receptacles like those used for carrying grape juice.' At about the same time Francis Bacon recognized the folly of applying farmyard manure and leaving it exposed to the sun – 'The ordering of Dung is, if the Ground be Arable, to spread it immediately before Plowing [sic] and Sowing and so to Plough it in. For if you spread it long before, the Sun will draw out much of the Fatness of the Dung: If the Ground be Grazing Ground, to spread it somewhat late towards Winter so that the Sun may have less power to dry it up.' It is sad to think that this four-centuries-old wisdom is often ignored today.

Composts prepared from dung, stalks, leaves, straw, weeds and other trash were widely used by the Greeks and Romans about the time of Christ. They are rarely used in modern agriculture because of the costs of making and storing them.

Untreated crop residues including cereal straw and stubble must also be regarded as manures with both long- and short-term benefits. The use of crop residues as manures reaches its ultimate in green manuring – that is the growing of a crop simply to incorporate it in the soil as a manure. None of it is harvested. Green manures are a most valuable source of nutrient elements and of organic matter. Green manuring is, like the use of farmyard manures, an ancient process and Pliny the Elder, in the first century AD, wrote 'It is universally agreed by all writers [!] that there is nothing more beneficial than to dig in a crop of lupins, *before they have formed pods*, using either the plough or the fork or also to cut them and bury them near the roots of trees or vines. It is thought also that where no cattle are kept it is useful to manure the soil with stubble or even fern.'

The choice of a leguminous crop, lupin, could not have happened by chance. Although they could not appreciate the nitrogen-fixing capacity of legumes the Romans were obviously aware of their superiority over non-leguminous crops as green manures.

In modern agriculture the former practice of setting aside one year in four or five for growing green manure crops to smother weeds and improve fertility has disappeared and green manure crops must now be fitted in to farming systems by using rapidly growing species in the intervals of a few months between economic crops.

Fertilizers

The term 'fertilizer' is used in this book to describe materials, mainly inorganic and commonly synthetic, which are rich in one or more of the essential plant nutrients.

Possibly the first fertilizers used were the 'marls', chalk and limestones used by the Romans about the time of Christ. The term 'marl' seems to have been applied to a wide range of earthy materials but they were probably all calcareous – rich in calcium carbonate – and this was the active constituent. Pliny the Elder, in the first century AD, said that 'It has been found in recent times that the olive thrives more particularly in soil that has been manured with the ashes of the lime kiln. Applications of chalk or limestone are particularly beneficial to the olive and the vine. When chalk is applied to the land the good effects may last for 80 years.' To do this the applications must have been massive and given grave risks of trace element deficiencies (mostly unidentified until this century).

By the Middle Ages liming and marling were well established in Great Britain. The marls used were 'limy clays' carted on to lighter, more acidic soils to improve both the physical and chemical properties. Local chalk, limestone or shell sand was also widely used. Gradually the importance of liming in the control of soil acidity was appreciated and liming is now an essential part of farming systems in areas with fundamentally acidic soils.

Developments in the use of other fertilizers were very slow until the emergence of modern agricultural chemistry in the nineteenth century. Before that the struggle towards an understanding of plant nutrition was painful indeed, with many wild theories being adopted and discarded. Practice preceded science in the use of simple fertilizer materials such as powdered bones and wood ash. They could be seen to improve yields but there was no under-standing of the reasons for this.

In the late sixteenth and early seventeenth centuries Francis Bacon experimented on the effects of a wide range of potential fertilizer materials on the germination and growth of wheat. His results showed that 'the most lusty, highest and thickest [wheat] . . . was produced by (1) Urine, (2) Dung, (3) Chalk, (4) Soot, (5) Ashes [presumably wood-ash], (6) Salt.' Wheat watered twice a day with hot water or treated with claret wine (!) did less well than his 'control' treatment – water only. He added that the wheat treated with Malmsey wine and Spirit of Wine 'came not up at all'.

Painstaking research of this kind was undoubtedly valuable but was bound to continue on a hit-and-miss basis until the surge of chemical knowledge in the eighteenth and early nineteenth centuries. During that period workers such as Sprengel, Boussingault and Liebig came to recognize that individual elements or their derivatives could affect the growth and composi-tion of the plant. This led to the birth of fertilizers as we know them today.

Most modern fertilizers supply nutrients in water-soluble forms to ensure rapid availability to the first crop after application. They include a wide range of 'compound' fertilizers which provide readily available nitrogenous compounds – nitrates, ammonium salts and urea, water-soluble phosphates and potassium salts such as potassium chloride, still known by the ancient name 'muriate of potash'. Water-soluble fertilizers are a relatively recent introduction, dating from the treatment of bones by sulphuric acid suggested by Escher in 1835 which led to the production of superphosphate by Gilbert and Lawes in 1843. This was the first 'artificial' water-soluble phosphate fertilizer and its use brought about huge yield increases of crops grown on phosphorus-deficient soils. It dominated the market as a phosphate fertilizer for over a century and was superseded only in the 1950s. It was also in the nineteenth century that ammonium sulphate was produced as a by-product of the coal-gas industry and potassium salts from Stassfurt were first used as fertilizers.

Superphosphate, ammonium sulphate and potassium chloride used individually are examples of 'simple' or 'straight' fertilizers. They were designed to supply *one* nutrient element only – N, P and K respectively. In fact they all supply other essential nutrients, sulphur from superphosphate and ammonium sulphate and chlorine from potassium chloride, but when they were first used these elements were regarded as incidental.

The same three substances – superphosphate, ammonium sulphate and potassium chloride – also formed the main basis of 'compound' fertilizers developed in the first half of this century. At first they were simply mixed together in varying proportions in powder form but during the 1930s similar mixtures were produced as soft rounded granules 2–3 mm in size. This was a major development and greatly improved keeping and spreading properties.

More recently, between 1950 and the present day, fertilizer manufacturers have greatly increased the concentration of N, P and K in their products by the introduction of such substances as ammonium nitrate, urea, and ammonium phosphates. As a result the present-day solid compound fertilizer consists of beautifully produced, free-running granules very rich in N, P and K.

'Liquid' or 'fluid' fertilizers have developed alongside the solid fertilizers which at present dominate the British market. They consist of solutions or more recently of suspensions of fertilizer salts, most of which are also used in solid fertilizers. Ammonia gas, liquefied under pressure and injected beneath the surface of the soil, is also used as a very concentrated source of available nitrogen.

Inorganic substances used to supply one or more trace elements to the plant also fall within the definition of fertilizers used in this book. The gradual recognition of the role of trace elements in plant nutrition, through the first recognition of deficiency symptoms in crops to the intimate biochemical functions of specific elements, has occurred almost entirely in this century. In fact Gris had discovered as early as 1844 that iron was essential for the correction of yellowing in vine leaves and the symptoms of manganese deficiency in oats were described by Salm–Horstmar in 1851. One by one the essential trace elements were identified – iron, manganese, zinc, boron, copper, molybdenum, chlorine, cobalt (for legumes) and others of no practical importance.

The need for fertilizers, manures and lime

The need for fertilizers and manures to support our present levels of cropping is amply evident on many of our soils if a small area is missed during application. Areas purposely left free of farmyard manure or accidentally devoid of fertilizer because of a blocked distributor spout in a placement machine or a gap between bouts show up rapidly. The plants lack vigour and are commonly pale green or yellow. This is usually a sign of nitrogen deficiency. Lack of most other essential elements is generally more slowly reflected in leaf symptoms. None the less the vigour of the plant will be affected by the lack of *any* essential element and crops may benefit greatly from nutrients added in manures or fertilizers despite the fact that no visual symptoms are shown by the untreated crop.

Many years of experimentation have shown that the greatest effects of fertilizer nutrients on crop yields have been brought about by nitrogen, phosphorus and potassium. As a result the modern fertilizer industry is based largely on supplying these elements in rapidly available forms.

As shown in Fig. 1.1, the increase in wheat yields over the past thirty years is very closely related to the rate of application of nitrogenous fertilizers but it must not be concluded that fertilizers alone have been responsible.

The use of NPK fertilizers *has* brought about very large increases in the yield of crops but this could not have been achieved without the parallel improvements in weed control, disease and pest control, the use of growth regulators and, above all, plant breeding. The outstanding example of the plant breeders' contribution has been in the production of cereal varieties which will not only tolerate but will respond, in terms of yield, to very much

Figure 1.1   *Average wheat yields and average use of nitrogenous fertilizers in the United Kingdom (1950–80). Data derived from the annual crop yield statistics of MAFF and the fertilizer statistics of the Fertilizer Manufacturers' Association.*

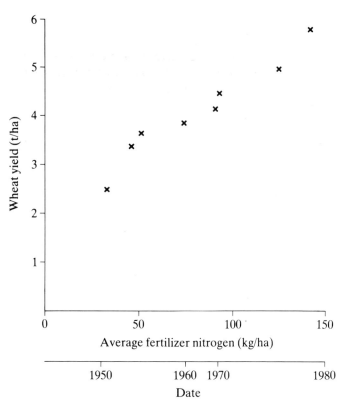

greater amounts of fertilizer nitrogen than was the case thirty years ago.

The progressively greater concentration of nitrogen, phosphorus and potassium in fertilizers has been achieved by the virtual elimination of the other major elements, sulphur, magnesium and calcium. NPK compounds, unless specially supplemented, also contain only very small quantities of the trace elements.

As a result the requirements of crops for essential elements other than N, P and K have had to be met from the reserves of nutrients in the soil, by contributions from liming materials (calcium and magnesium), from pollution (mainly sulphur) and from manures (all elements). In many cases, these contributions have been insufficient to meet the needs of the much heavier crops produced under the influence of NPK fertilizers. Further, the rise in soil pH values, brought about by liming, has reduced the availability of magnesium and most trace elements to the plant. By a strange paradox in these days of sulphur-laden acid rain, pollution is insufficient in many areas to supply the needs of crops now that little or no sulphur is present in fertilizers.

All these factors have combined to increase the strains on the nutrient resources of soils, especially under intensive, high-yield systems; deficiencies of magnesium, sulphur, manganese, boron and copper are much more prevalent than they were thirty years ago.

Opponents of fertilizer use raise many criticisms of the effects of fertilizers as 'poisons' affecting, above all, the quality of food. An often-quoted example is the presence of nitrates in baby foods. Such worries are not surprising and are occasionally well based in fact. In such cases there has usually been gross abuse of fertilizers. If used incorrectly or in gross excess, some fertilizers *can* adversely affect germination, damage growing crops, reduce yields as compared with more modest dressings and cause stream pollution (especially by nitrates). Such fertilizer practices are both in-efficient and inexcusable and one of the aims of this book is to show why and how they should be avoided.

If fertilizers are used properly, the nutrients they contain become virtually indistinguishable within a few days of application from what was already in the soil. Slowly soluble fertilizers such as lime, gypsum and ground mineral phosphate simply supplement the reserves of calcium, magnesium, phosphorus and sulphur in forms already present in many soils. Most of the nutrients in water-soluble fertilizers are rapidly – almost greedily – assimilated by the soil and its organisms. For example, water-soluble phosphates are sorbed by the soil solids within a few hours, although they will remain readily available to plants for some months during which they are being changed by the soil into less available forms. The fertilizer cations potassium (K^+) and ammonium (NH_4^+) are also rapidly adsorbed by clay and humus and are thereafter impossible to distinguish from potassium and ammonium ions already on the clay/humus complex unless (and this is done only for research purposes) they are 'labelled' with isotopes. Fertilizer urea is rapidly converted by the soil, in the same way as urea from animal excreta, to ammonium nitrogen and thereafter acts as such. Fertile soils are already rich in nitrates and fertilizer nitrate simply improves the fertility of soils less rich. The only component of most NPK fertilizers, applied at 'normal rates', which greatly exceeds the requirements of the crop is the chloride ion derived from potassium chloride used to supply the 'K' component.

Thus within a few days of fertilizer application, provided there is sufficient water to dissolve the nutrients, the soil should be almost identical to a soil of similar parent material and physical properties which has received a good supply of nutrients from farmyard manures or green manures ploughed in. The idea of fertilizers as 'poisons' must be rejected.

The rôle of fertilizers as pollutants must be taken much more seriously. The risk of substances being leached from the soil in amounts likely to cause harm to aquatic animals or to human beings is much lower than the risks from slurry, farmyard manure or silage effluents entering watercourses. The 'biological oxygen demand' (BOD) of fertilizer leachates, in contrast to that of manure effluents, is negligible.

The real problem with fertilizer leachates concerns nitrates. The soil has no mechanism for retaining nitrate ions. If there is sufficient rainfall, and the soil is well drained, fertilizer nitrates will move downwards through the soil and, unless caught up by plant roots, will enter the drains. Ammonium nitrate is a major component of both simple and compound fertilizers. Half of its nitrogen is in nitrate form. In a very wet spring such as 1983 a large proportion of this nitrate will be leached, especially if applied to arable crops at or before sowing. There was ample evidence of this in both the yellowing of crops brought about by lack of nitrogen and in nitrate concentrations in watercourses.

Such evidence suggests that in the very wet season of 1983 some 20–30 per cent of applied nitrogen was leached. On a UK basis this would represent a loss of 300 000–400 000 tonnes of fertilizer nitrogen worth £120–160 million, with consequent stream and river pollution. The year 1983 was abnormal, but in a year of *average* rainfall it would be realistic to assume a leaching loss of 10–20 per cent of fertilizer nitrogen from arable cropping areas with smaller losses of less than 10 per cent from established grassland. Even this would represent £40–60 million worth of fertilizer on a UK basis.

The use of lime and manures has been established for many centuries and their effects are relatively well understood. Fertilizers, as we know them today, have been produced and developed much more recently and our understanding of the factors affecting the choice of types and amounts of fertilizer to use for a particular crop are far from complete. One of the aims of this book is to summarize the present state of knowledge and to assess how best to use it.

Chapter 2 The nutrition of plants

In order to grow efficiently, plants when germinated need light, air, water, warmth and a range of plant nutrients. In agriculture these factors are governed by the complex interactions of the soil, climate and atmosphere.

The main source of plant nutrients is soil. In areas of natural vegetation, uncultivated and not greatly affected by man's activities, the types and numbers of plants which survive are strongly influenced by the nutrients that can be obtained from the soil and, indirectly, from the atmosphere. A soil which is deficient in available phosphorus will support only pine forest or heather as on the strongly acid hills of Scotland, or poor grass-heath as on the alkaline chalk downlands of southern England. The natural vegetation over large areas of the British Isles before agriculture was forest – coniferous in less fertile acidic areas and broadleaved in more fertile areas of mildly acidic or alkaline soil. Most of this natural vegetation has now been replaced by arable crops or grassland and only remnants of the forests now exist.

In agriculture, crops are grown which are not necessarily natural to the soil. The constant aim is to increase the yield of crops and the continual removal of high-yield crops puts a strain on the ability of the soil to supply sufficient nutrients and brings about the need to use lime, manures and fertilizers.

Plants which are grown with adequate light, heat, atmospheric oxygen and carbon dioxide and sufficient but not excesive amounts of water and plant nutrients will develop a strong root system which will provide a firm anchorage and a capacity to forage for further nutrients. With these provisions green plants grow by creating or synthesizing carbohydrates, fats, proteins and nucleo-proteins. These substances are composed mainly of carbon (C), hydrogen (H), oxygen (O), nitrogen (N), phosphorus (P) and sulphur (S), and a very large proportion of the plant consists of these essential elements. For healthy growth the plant needs several other elements to maintain the osmotic pressure, to form

part of tissue material or to act as catalysts in some of the
thousands of enzymic reactions involved in plant growth. A list of
elements known to be essential to plant growth is given in Table
2.1, along with examples of their functions.

Table 2.1   Elements essential for plant growth.

Major elements		
Element	Symbol	Examples of functions in the plant
Carbon	C	Component of all carbohydrates, fats, proteins. Essential as carbon dioxide in photosynthesis.
Hydrogen	H	Component of organic matter. Essential as a component of water to photosynthesis and other biochemical processes.
Oxygen	O	Component of organic matter. Essential as a component of water and carbon dioxide to photosynthesis. Essential for respiration.
Nitrogen	N	Component of all proteins and amino-acids.
Phosphorus	P	Constituent of cell nucleus. Essential for cell division. Key to many enzyme actions.
Potassium	K	Osmotic pressure regulation. Transfer of carbohydrates within the plant. Role in enzyme actions especially in protein synthesis.
Sulphur	S	Component of some proteins, amino-acids and oils.
Calcium	Ca	Enzyme activation. Component of calcium pectate in cell walls. Essential for cell division. Regulation of osmotic pressure.
Magnesium	Mg	Constituent of chlorophyll. Enzyme activation. Regulation of osmotic pressure.
Trace elements		
Manganese	Mn	
Iron	Fe	
Boron	B	Many specific functions mainly in enzyme systems essential to photosynthesis, nitrogen assimilation or protein formation.
Copper	Cu	
Zinc	Zn	
Molybdenum	Mo	
Cobalt	Co	
Chlorine	Cl	

The elements in Table 2.1 have been divided into 'major' elements – those required by the plant in relatively large quantities, conveniently measured in grams of nutrient per kg of plant dry matter, and 'trace' elements – required in much smaller quantities, conveniently expressed as mg of nutrient per kg of plant dry matter. The two groups are quite distinct. The actual amounts of individual nutrients required will vary from crop to crop, but a trace element for one species will never be a major element for another species and vice versa.

It is very important to realize that the terms 'major' and 'trace' refer to the *amounts* of the elements required by plants and do not imply major and minor importance in plant nutrition. The terms are misleading in some ways. Plant growth is dependent upon a combination of *all* the essential elements and the lack of any one of them, whether required in very large amounts, such as nitrogen or potassium, or in very small amounts, such as molybdenum or copper, will restrict growth. For example, a root crop such as turnip or mangold may need more than 100 kg/ha of nitrogen, 20–30 kg/ha of sulphur and only 0.1–0.2 kg/ha of copper, but the availability of any less than these quantities to the crop will restrict yield.

There are many published works showing the variations in the amounts of individual elements taken from the soil by different crops. By and large they show that cereal crops need less mineral nutrients per hectare than any other crops. In contrast grass, clover and root crops take up much more. The demand of the plant for a particular nutrient is not constant from week to week. There is usually less need for nitrogen as a plant matures but crops such as regularly-cut grass, which are not allowed to mature, have a high nitrogen requirement throughout the season. Most arable crops have a peak period of nutrient requirement when they are growing most vigorously, as for example in the period of rapid haulm growth in potatoes (Fig. 2.1). Thus the *rate* at which the soil can supply available nutrients may become restricting to crop growth. This is why the timing of application of fertilizers, methods of application such as placement or top dressing and the possible use of slow-release fertilizers become important – to ensure maximum availability at the time of maximum requirement.

It has become accepted that plant *nutrients* are given the names of the *elements* primarily involved. We speak of phosphorus as a plant nutrient although it is always taken up by the plant in the

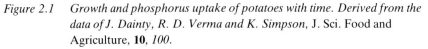

Figure 2.1   Growth and phosphorus uptake of potatoes with time. Derived from the data of J. Dainty, R. D. Verma and K. Simpson, J. Sci. Food and Agriculture, **10,** *100.*

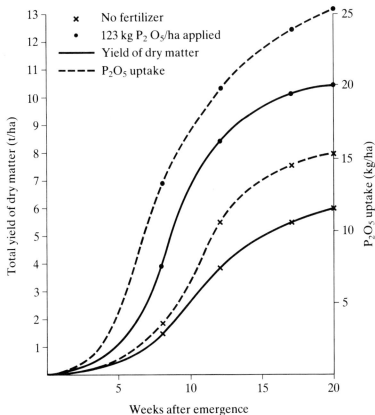

form of phosphates. Almost all nutrients are absorbed by the plant either as simple chemical compounds or as ions. For example, carbon, hydrogen and oxygen enter the plant as carbon dioxide (CO_2) through the leaves or as water (H_2O) mainly through the roots.

All other nutrients enter the plant mainly through the roots, either as positively charged *cations* such as ammonium (NH_4^+), calcium (Ca^{2+}), potassium (K^+) or as negatively charged *anions* such as nitrate (NO_3^-), sulphate (SO_4^{2-}) or phosphates ($H_2PO_4^-$, HPO_4^{2-}). It is in the form of such ions that mineral nutrients become available to the plant.

The rôles of individual elements in plant nutrition

Carbon, hydrogen, oxygen

These major elements must be regarded separately from the others. They enter the plant mainly in non-ionic forms as carbon dioxide, water and oxygen. Provided that a plant is growing well with adequate supplies of other essential nutrients and adequate light, the intake of these elements by the plant will occur automatically.

Nitrogen

Nitrogen is contained in all proteins and nucleic acids and in all protoplasm. It is taken up by the plant, mainly through its roots, as ammonium ions (NH_4^+) or as nitrate ions (NO_3^-). In the plant the nitrate is rapidly converted to ammonium which combines with carbohydrates formed during photosynthesis to form amino-acids and, eventually, proteins. The protein formed causes the leaves to grow and increase their green surface area, thus increasing photosynthesis and stimulating further growth. In many crops the total leaf area is proportional to the nitrogen supply. The relationship breaks down only if nitrogen is absorbed in excessive amounts which cannot be used by the plant in protein synthesis. This excess nitrogen performs no useful function and remains as non-protein nitrogen, some of it in the form of nitrate.

As the nitrogen supply increases, the proportion of protein to cell-wall material (mainly carbohydrates) is also increased. This gives more succulent leaves containing more water. If the nitrogen supply is excessive the leaf cells become very large and thin walled and the leaves are easily injured by wind, rain, frost, fungi or insects. Because of their high nitrogen content the leaves also make ideal food for bacteria which invade tissue damaged by wind or mechanical means and multiply rapidly.

Lodging or flattening of cereal crops, vividly seen where fertilizer overlaps have occurred, is a direct and often costly consequence of the weakening of stem cell structure by excess nitrogen.

Nitrogen-deficient plants are dwarfed and have small pale yellow leaves containing little chlorophyll. If the supply of available nitrogen is increased there is usually a very rapid response in leaf size and chlorophyll content: leaves become a darker green. Leaves with excess nitrogen are deep blue-green and soft. Excess nitrogen also tends to keep the leaves green longer and thus to delay maturity.

Less attention has been paid to the effects of nitrogen on stem and root tissue but there is good evidence that both are increased roughly in proportion to the increase in leaf yield. The effects of nitrogen on plant roots have been particularly neglected until recent years and many textbooks quote phosphorus as the main element affecting root growth. This is not true. The effects of nitrogen on the elongation and ramification of roots in soil are very important and greatly affect the ability of the root system to forage for other nutrients within the soil. This can be seen in the effects of top dressings of nitrogenous fertilizers on yellowing crops with restricted root systems whether resulting from nitrogen starvation or from other causes such as temporary waterlogging or soil compaction. The extra nitrogen will not solve the basic problem but can help the plant to create a surface root system sufficient to save the crop.

The effects of increasing nitrogen nutrition on the yield of all parts of the plant, and particularly on the harvestable yield of cereals, roots, grass and potatoes, are greater than for any other nutrient. Accurate assessment of the nitrogen requirements of crops at different stages of growth and the supply of sufficient but not excess nitrogen are critical to optimum crop production.

Phosphorus

Phosphorus is taken up by the plant roots mainly as ortho-phosphate ions ($H_2PO_4^-$, HPO_4^{2-}). It has an absolutely vital rôle in the plant, being part of the cell nucleus, essential for cell division and, therefore, particularly important at the growing points of the plant, the meristematic tissue. The autoradiography technique, used on plants treated with radioactive phosphates, indicates on a photograph the extraordinary concentration of phosphorus at the growing points of leaves and roots. The concentration of phosphorus in dividing cells can be more than a thousand times that in mature cells.

Phosphorus is also essential to the plant because of its rôle in many plant enzyme actions, for instance during the fascinating and complex conversion of water and carbon dioxide to sugars and starches in the process known as photosynthesis.

Because of restricted cell division plants deficient in phosphorus cannot grow. The main symptom of phosphorus deficiency is stunting of leaf, stem and root systems. In the field, the result can be a bizarre picture of miniature plants, sometimes with purplish leaves. The purple colours have been quoted as being specific symptoms of phosphorus deficiency. They are not. They can result from numerous other causes – waterlogging, drought, magnesium

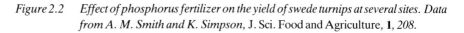

Figure 2.2   *Effect of phosphorus fertilizer on the yield of swede turnips at several sites. Data from A. M. Smith and K. Simpson*, J. Sci. Food and Agriculture, **1**, *208.*

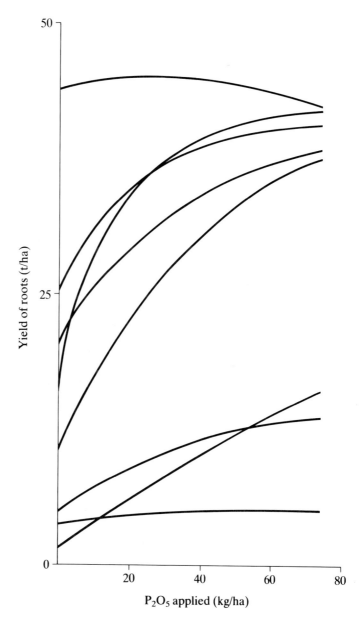

deficiency and nematode attack included – as well as phosphorus deficiency.

Another statement, passed down from generation to generation, that phosphorus has a *specific* effect on root development, cannot be supported. While it is true that phosphorus is vital for the growing tips of roots, other essential elements, particularly nitrogen and calcium, are very much involved in good root growth. Deficiency of phosphorus will affect all parts of the plant as phosphorus has many other functions besides that in meristem tissue.

The response of crops grown on phosphorus-deficient soil to an increase in phosphate supply is quite remarkable, as is illustrated in Fig. 2.2 which shows the results of field experiments on a swede turnip crop made on mainly phosphorus-deficient soils. The only exception is shown in the top graph in which fertilizer phosphorus had little effect on yield because the supply from the soil alone was sufficient for the crop. The very flat graph at the bottom of the series demonstrates the limiting effects of other factors on the capability of a crop to respond to a nutrient. In this case yield was severely restricted by soil acidity.

Potassium

Unlike the other major elements, potassium is not a component of proteins, carbohydrates or any other main substances in the plant. It is readily absorbed by the plant roots as the potassium ion (K^+) and this is retained mainly in the cell sap, playing a part in regulating osmotic pressure and maintaining the turgidity of the plant.

Potassium is also involved in the essential processes of photosynthesis and respiration as well as the movement of carbohydrates from one part of the plant to another. Examples of this are the transfer of sugars from the leaves to the swelling roots of sugar beet where they accumulate as sucrose, to the tubers of potatoes where they are converted mainly to starch or to the seeds of cereals, peas and beans where they are converted to starch and protein. Although it is not a component of proteins there is now strong evidence that potassium is an activator of the essential enzyme actions involved in protein synthesis.

Potassium moves very freely from soil to plant and also within the plant. As a result luxury uptake of potassium can occur if there is a large supply of potassium available from the soil. This can interfere with the uptake of other ions, notably magnesium, causing magnesium deficiency in the plant.

If there is a danger of potassium deficiency the plant is capable of transferring potassium from older to younger leaves in its efforts to survive. This upsets the osmotic balance in the older leaves which lose water from their margins and between the main veins. These areas become 'scorched' and brown in appearance and give the typical symptoms of potassium deficiency. Similar symptoms can be seen on older leaves when normal plants become senescent. As the plant matures and leaf cells start to decay much of the potassium is leached from the leaves and returns to the soil. This occurs much more readily with potassium than with other essential elements which become part of complex substances in the plant.

Sodium

Some of the functions of potassium in the plant, particularly in the regulation of osmotic pressure and maintenance of turgidity, can be performed by sodium. It is not, however, truly an essential element for most crops as they can survive without it. Also it cannot replace potassium in its specific enzyme-activating functions. The partial replacement of potassium by sodium can be made in most crop species although in practice its use is restricted to sugar beet, mangolds and occasionally to vegetable crops.

Sulphur

Sulphur is absorbed by plant roots mainly as the sulphate ion (SO_4^{2-}). Sulphide ions (S^{2-}) formed under anaerobic conditions in soil may also be absorbed but are toxic to plants.

Sulphur is an essential component of many proteins, amino-acids and some vitamins including biotin and thiamine. Sulphur is also contained in the oils produced by plants of the *Brassica* family, such as kale and oil-seed rape. Thus brassica crops have a high sulphur requirement and are much more susceptible to sulphur deficiency than groups of crops such as cereals which need less sulphur. The symptoms of sulphur deficiency, yellowing of the leaves and dwarfing of the plant, are difficult to distinguish from those of nitrogen deficiency.

Calcium

Calcium is taken up by plant roots as the calcium ion (Ca^{2+}) and has a function in osmotic pressure regulation. Like phosphorus it is essential at the growing points of plants, especially root tips. It occurs in cell-wall materials as calcium pectate and also plays a part in preventing the accumulation of excessive and toxic amounts of manganese in the plant. Calcium deficiency occurs mainly in very acidic conditions and symptoms shown by the plant – short, stubby, scorched root systems, stunted stems and leaves –

are usually the result of toxicity of manganese or aluminium. Specific symptoms of calcium deficiency are uncommon. They include cupping of the leaves of brassica plants. Calcium deficiency may also occur in rapidly growing crops even when the supplies from the soil seem to be ample. This is apparently the result of the relatively slow movement of calcium in the plant and can result in 'tip-burn', a scorching of the leaf tips, especially in rapid growing horticultural crops such as lettuce.

Many plant species are calcifuge – they do not grow well in soils containing much available calcium – including a number of prized horticultural species such as heathers. They react adversely both to high soil pH and to high calcium levels in the soil and suffer easily from manganese and magnesium deficiency. It is not within the scope of this book to discuss the mechanisms involved, as no temperate agricultural crops can be classed as calcifuge. The nearest approach is in crops such as oats, rye and potatoes which prefer acidic soils but none the less require adequate calcium nutrition.

Magnesium

Magnesium is taken up by plant roots as magnesium ions (Mg^{2+}). It plays a part in the regulation of osmotic pressure but is an essential element mainly because it is contained in the chlorophyll molecule. It also assists in the necessary and rapid movement of phosphorus within the plant.

Plants which are acutely deficient in magnesium cannot form chlorophyll. Magnesium deficiency symptoms usually take the form of blotchy yellowing followed by browning of the leaves. The areas around the main veins and the leaf margins do form some chlorophyll and remain green longest.

Trace elements

Trace elements known to be essential for plant growth are manganese (Mn), copper (Cu), boron (B), iron (Fe), zinc (Zn), molybdenum (Mo) and chlorine (Cl). In addition, cobalt (Co) is needed by many micro-organisms including the nitrogen-fixing bacteria, *Rhizobia*, which live in the root nodules of legumes and supply the plants with nitrogen. Cobalt is therefore essential to the efficient growth of leguminous crops such as peas, beans, clovers and lucerne.

There is also evidence that sodium (Na), aluminium (Al), silicon (Si), selenium (Se) and vanadium (V) are essential at least for some species of plants but recorded cases of deficiency of these elements are very rare.

All trace elements have very specific functions in plant nutrition, most of which are concerned with enzymes and co-enzymes. Enzymes are proteins of many kinds, produced by living plant cells, which act as catalysts to many of the biochemical reactions on which healthy growth depends. The enzymes act along with non-protein substances known as co-enzymes. As an example, iron is known to take part in thirty or more enzyme systems in plants. Some of the 'iron proteins' involved have very large molecules which contain only one or two atoms of iron. None the less this 'trace' of iron is critical to the enzyme systems.

Typically, only very small amounts of trace elements are required to perform their special but essential functions. Also, during their catalytic action the enzyme molecules are not permanently changed but can be re-used.

Table 2.1 illustrates some of the processes for which trace elements are essential.

Chapter 3 Nutrient requirements of crops

Requirements for individual nutrients vary considerably from crop to crop and, more important, from one part of the growth period to another. Table 3.1 gives some examples of the approximate quantities of the major elements taken from the field at harvest by various crops at given yields. They are compared with the amounts taken up by a natural grass-heath sward growing on an acid soil, pH 5.0. The range of trace element uptakes by crops is included at the foot of the table.

From the data in Table 3.1 it is clear that the demands upon nutrients are much greater in the high-yielding arable crops or intensive grass than in the natural grass. Hence there is a need in productive agriculture to supplement the amounts of nutrients that the soil can supply. This is done by the use of manures, lime and

Table 3.1 Approximate amounts of nutrients (kg/ha) removed annually in crops.

Crop	Yield/ha	Dry matter yield/ha	N	P_2O_5	K_2O	Ca	Mg	S
Cereal	6 t grain 3.5 t straw	8 t	120	50	70	15	10	30
Sugar beet	40 t roots 25 t tops	12 t	200	45	240	70	25	30
Potatoes (tubers only)	50 t	10 t	180	50	240	10	15	20
Grass silage	30 t	—	160	40	160	45	15	15
Hay	—	8 t	100	30	120	30	10	10
Clover	—	5 t	180	25	120	100	15	15
Kale	50 t	10 t	200	60	220	250	20	100
Natural grass heath	2 t	0.4 t	10	3	10	2	1	2

Trace elements (g/ha) removed annually in crops

Iron (Fe)	600–2 000	Zinc (Zn)	100–400
Manganese (Mn)	300–1 000	Copper (Cu)	30–100
Boron (B)	50–300	Molybdenum (Mo)	5–20

Source: Data from several sources.

fertilizers. Most commercial fertilizers supply only nitrogen, phosphorus and potassium. Most liming materials supply mainly calcium, but some contain calcium and magnesium in roughly equivalent amounts. The British Isles in common with many industrial countries have relied for several generations on sulphur in pollution, produced by burning coal or oil products, to support crops. Little sulphur is added in modern fertilizers and only manures help to supplement soil sulphur supplies. Although the amounts of magnesium and sulphur taken up by crops are similar to, and sometimes greater than, the amount of phosphorus, they are commonly neglected in fertilizer practice whereas phosphorus is used in large amounts.

Values of 'offtake' of nutrients in the harvested crops as shown in Table 3.1 are useful in *comparing* the needs of various crops and in assessing the final annual loss of nutrients from the soil. They can, however, greatly underestimate the amount of a nutrient required to produce the crop. One reason for this is that published data commonly ignore that part of the crop which is not harvested – cereal roots and stubble, potato haulms. Another is that losses of some nutrients, especially potassium, occur by leaching from the

Figure 3.1   *Accumulation and loss of N, P_2O_5 and K_2O in winter wheat with time. Adapted from the data of F. Knowles and J. E. Watkin, J. Agric. Sci., **21**, 612.*

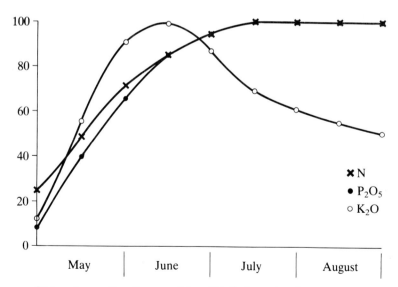

✗ N
● P_2O_5
○ K_2O

*At each sampling the quantities of N, P_2O_5 and K_2O in the above-ground parts of the crop are expressed as a percentage of the greatest quantity present at any of the samplings

plant tissues as the crop ripens. This is illustrated in Fig. 3.1 in terms of percentage of maximum uptake.

A further problem in assessing the requirement of a crop for a particular nutrient arises from the very variable demands of the crop from week to week during the growing season. Most single-harvest crops (cereals, potatoes, sugar beet) have a growth pattern which follows an S-shaped curve with time. This is shown in Fig. 2.1 in terms of the amount of dry matter accumulated during each week of growth. The crop grows very slowly for some weeks and much of the early growth is in the root system. Potatoes, for example, can produce as little as 3–5 per cent of their total dry matter yield in the first six to eight weeks of growth. There is then a remarkable surge of stem and leaf growth which may even be observed from day to day. Then, as the crop matures, the amount of dry matter actually accumulated by the plant per week becomes quite small, although vigorous changes may be taking place within the plant, transferring materials from leaf to grain, root or tuber.

The demand for nutrients to service plant growth during the middle period is very great and success or failure in meeting the demand is a large factor in determining the final yield of the crop. It is met partly by a gradual build-up of nutrients in the plant during the initial slow-growth period but there is a very critical phase at and just before the growth flush starts when the uptake of elements such as nitrogen and potassium must be very rapid. In fact, as shown for the potato crop in Fig. 2.1 (p. 14) the plant anticipates its grand period of 'top' growth by an earlier rapid increase in phosphorus uptake. This is also the case for both nitrogen and potassium. More than half the plant's total uptake of nitrogen can occur in the first quarter of the growth period (Fig. 3.1) and, by the half-way stage, it had taken up 80 per cent of its total requirement. This pattern of growth and uptake is typical of most crop species. There is clearly a period of maximum stress on soil nutrient reserves just before and during the period of rapid growth.

Against this must be set the fact that during the initial slow-growth phase healthy plants will have established a vigorous, well-ramified root system capable of tapping a large volume of soil for water and nutrients. It follows that any restriction of root growth during this early period caused by poor soil structure, pans, waterlogging or inadequate nutrient supply may seriously affect final yield.

Responses of crops to nutrients

The response of a crop, in terms of total dry matter yield, to a single nutrient element can be measured in pot or field experiments by using increasing rates of that nutrient in statistically designed experiments and measuring the yields produced. The conditions required for success in such experiments are very important. We have seen that no single nutrient can be fully effective in increasing crop yields unless all other essential elements are available in sufficient quantities that *they* do not restrict yield. Also, in order to avoid antagonistic effects they must not be present in excessive amounts. The easiest way to achieve this is in pot experiments using inert growth media such as pure silica sand instead of soil and supplying the nutrients as solutions of salts containing them. Thus, for example, nitrogen, potassium, sulphur, calcium, magnesium and all the trace elements may be added to a growth medium, leaving only phosphorus out of the basal nutrient solution. The specific effect of phosphorus on yield can then be measured by regulating the amount of phosphorus added as phosphate to individual pots.

This is described by the Law of the Minimum first put forward in 1843 by Liebig, which states that 'Plant growth is regulated by the factor present in minimum amount and rises or falls according as this is increased or decreased in amounts.'

Unfortunately many other factors other than nutrient supply can restrict yield according to the Law of the Minimum and in order to test the effect of a nutrient *in ideal conditions*, light, temperature, air movement, humidity and water availability must also be ideal. This can be achieved in carefully controlled experiments in growth chambers. The results in terms of the *shape* of the yield curve are very similar for all nutrients. A typical yield curve, to the point of maximum yield, with increasing supply of any nutrient is shown in Fig. 3.2, Curve 1. The yield response when little of the nutrient is available is very steep. Further responses become smaller for each amount of nutrient added until a maximum yield is reached.

Such experiments give a measure of the *potential* effects of a nutrient on crop yield. Although they provide a goal to aim for in crop production such yields are not attainable in the field as, unfortunately, field conditions are far from ideal. Chemical reactions and physical features of the soil, such as poor structure, restrict the availability of the nutrient to the crop. Also, in climates similar to that of the British Isles there are great day-to-day and year-to-year variations in temperature, rainfall, wind speed and

Figure 3.2   Effect on crop yield of increasing the supply of a nutrient.

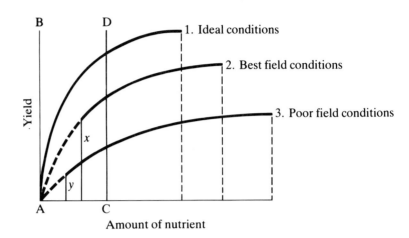

humidity. The result is that yield responses in agriculture follow patterns similar to those shown in Fig. 3.2, Curves 2 and 3. Curve 2 represents the best response obtainable in field conditions whereas Curve 3 would apply where severe restrictions to crop yield arise from other sources such as waterlogging or persistent low temperatures.

It is important to note that the *shapes* of Curves 1, 2 and 3 *are similar*: large responses at low levels of nutrient and smaller responses as more nutrient is supplied, eventually reaching a maximum. The differences between the curves are in:

● The maximum yield attained.
● The amount of nutrient required to give that maximum.
● The amount of yield increase brought about by each increase of nutrient supplied.
● The yield with no added nutrient. This is zero when, as in the 'ideal' experiments, no nutrient is supplied by the growth medium. Even the poorest of soils supply some nutrient, however small in quantity. This can be represented by moving the 'starting point' of the curve from line AB to line CD or some other parallel line depending on the amount of nutrient the crop is able to obtain from the soil. The basic yields obtainable from the soil without added nutrient are *x* (Curve 2) and *y* (Curve 3).

Thus the basic yield potential, the response to added nutrients and the amount of nutrient required to give maximum yield for a particular soil/climate vary greatly from area to area and even from field to field.

Correct assessment of the amounts of nutrients required and the responses expected together with an appreciation of the properties of fertilizers other than their nutrient content are the basis of good fertilizer and manuring policy. The practical aspects of this are discussed fully in Chapter 13.

Chapter 4 Factors affecting the availability of nutrients to plants

The concept of availability

Plants can absorb and use nutrients only if they are in simple forms, usually ionic. Soil minerals such as micas, feldspars, hornblende and augite are complex silicates containing among them potassium, sodium, calcium and magnesium but these elements are of no immediate value to the plant and are said to be 'unavailable'. They become 'available' by conversion through weathering of the rock mineral to ionic forms (K^+, Ca^{2+}, Mg^{2+}, Na^+) which enter the soil solution. All ions in the soil solution are available to the plant, provided they can reach an active root or it can reach them. The concentration of ions in the soil solution at any given moment is, however, usually low, except immediately after the application of water-soluble fertilizers.

Most nutrients are taken up by the plant in the form of cations which are positively charged. Available cations either in the soil solution or held against leaching by negatively-charged clay or humus particles are easily taken up by the plant.

Other plant nutrients, notably phosphorus, boron and molybdenum, are taken up by the plant as anions which are negatively charged. The concentration of anions in the soil solution is usually low because they become converted to new, less soluble chemical forms. Phosphate ions, for example, may be converted to insoluble iron or aluminium phosphates which are less taken up by to the plant.

Nutrients contained in soil organic matter are also available in varying degrees to the plant. The organic matter is derived from plant residues, manures and dead or alive soil animals including a vast population of micro-organisms. Substantial proportions of nutrients contained in organic matter are in complex forms unavailable to the plant. They include the nitrogen, phosphorus and sulphur contained in the persistent organic materials. Other organic matter, particularly from recent crop residues or manures, can be decomposed quickly by bacteria and thus 'mineralized'. Their carbon is released to the atmosphere as carbon dioxide and the nutrient elements become available to the plant as ammonium, nitrate, phosphate, sulphate and other ions.

In both inorganic and organic fractions of the soil the amount of a particular nutrient in forms chemically available to the plant varies from day to day and even from minute to minute if a vigorous crop is absorbing it. The system is in a continuous state of flux.

Figure 4.1   The availability cycle of nutrients.

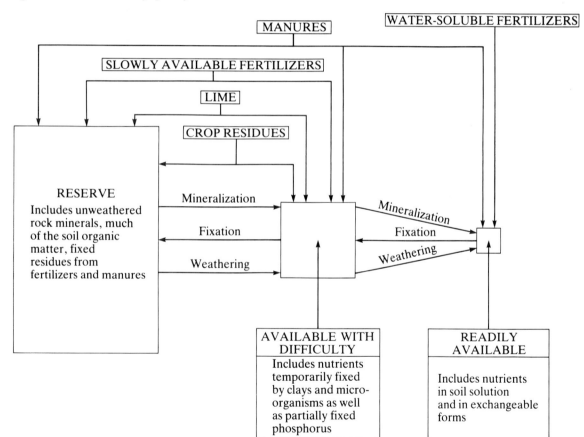

The ever-changing pattern is broadly represented in Fig. 4.1. The sizes of the 'boxes' indicate very roughly the proportions of a nutrient in three empirical fractions: 'reserve', 'available with difficulty' and 'readily available'. Taking potassium as an example the 'reserve' would include potassium locked up in rock minerals such as potash feldspars. 'Available with difficulty' would include potassium trapped temporarily in clay lattices and 'readily available' would include potassium ions in the soil solution or held on the clay and humus as 'exchangeable' potassium.

However much 'readily available' nutrient a soil may contain, various physical features of the soil may make it impossible for the plant roots to absorb it. It then becomes 'positionally unavailable'. These features include pans and other forms of poor soil structure, waterlogged or

Figure 4.2   Competition for nutrients in the soil.

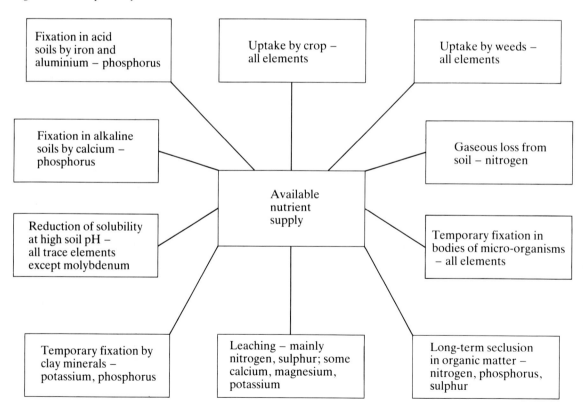

excessively dry horizons. Thus a soil might contain large amounts of available phosphorus within clods which roots cannot penetrate and from which the phosphorus can diffuse only very slowly. In such cases the so-called 'availability' of the nutrient becomes meaningless.

Competition for nutrients

Figure 4.2 illustrates the competition that a crop plant must overcome from inorganic components of the soil and from organic matter, living or dead.

Because of the many competing factors the proportion of the total plant nutrients in the soil taken up annually by the crop is very small, in some cases much less than 1 per cent. Even newly applied water-soluble nutrients from fertilizers are subject to leaching, uptake by weed species or other soil organisms, and fixation in forms less available to the plant. The result is, commonly, that only some 15–25 per cent of fertilizer phosphorus is taken up by the first crop and very much less by subsequent crops. The uptake of fertilizer nitrogen, especially by vigorously growing grass crops, can be much greater – in rare cases as much as 80–90 per cent of the amount applied.

In general, however, the crop plant is a poor competitor for nutrients from soil, manure or fertilizer.

Factors restricting availability

Table 4.1 summarizes the many factors which restrict nutrient availability and gives examples of the elements mainly affected. The right-hand column indicates the likely effectiveness of fertilizers or manures used to alleviate resultant deficiencies. The factors are presented in four groups – chemical, physical, biological and climatic – although there are considerable overlaps between the groups.

Table 4.1   Factors restricting the availability of nutrients to the plant.

Restricting factor	Effects on specific nutrients	Crop responses to supplements from manures or fertilizers

Chemical:
  Alkalinity
  (high soil pH)
	Phosphorus – fixed as tri-calcium phosphate in calcareous soils.
	Manganese, iron, zinc, boron, copper, cobalt – compounds are converted to less soluble forms.
	Nitrogen – some loss of ammonia by volatilization, especially after applications of urea or ammonium fertilizers.
	Potassium, magnesium – availability reduced by antagonism of excess calcium.
		Good short-term responses but added nutrients will be subject to the same adverse influences as nutrients already in the soil and will rapidly decline in availability. For trace elements, sprays applied to foliage are commonly more effective than applications to the soil.

Table 4.1   (Cont)

Restricting factor	Effects on specific nutrients	Crop responses to supplements from manures or fertilizers
Chemical: Acidity (low soil pH)	*Phosphorus* – fixed as iron and aluminium phosphates. *Nitrogen* – poor nitrification, poor nitrogen assimilation by soil bacteria; clovers, peas and beans (most legumes) cannot tolerate low pH. *Calcium, magnesium, potassium, copper, zinc* – strongly leached in acidic soils of high rainfall areas. *Molybdenum* – converted to unavailable forms. [NB Manganese and aluminium become more soluble and may become toxic to the plant, restricting the uptake of other nutrients.]	Response restricted in very acid soils because of damage to root systems, especially in sensitive crops such as barley and sugar beet. Liming essential to ensure more efficient uptake by the plant of both indigenous and added nutrients.
Antagonism of a gross excess of another element	Any nutrient cation may be affected. The most common examples are *magnesium* deficiency induced by excess potassium – usually derived from fertilizers or cattle slurry, and *potassium* or *magnesium* deficiency induced by excessive liming with calciferous limes.	Good short-term responses. The antagonisms can be greatly reduced by sensible policies for use of lime, fertilizers and manures – avoiding excessive applications.
Low cation-exchange capacity (sandy soils containing little humus)	All cations subject to leaching. Appreciable losses of *calcium, magnesium* and *potassium. Copper* and *zinc* may become deficient.	Good responses, but added nutrients are also subject to rapid leaching, especially if water soluble.
Absolute deficiency	*Calcium* – in acid soils. *Potassium* – in many sandy soils and deep peats. *Phosphorus* – some peats. *Copper* – in some reclaimed peats or podzols and in some calcareous soils.	Immediate and vigorous responses unless there is some severe restricting factor such as extreme acidity.
Fixation by clays	*Potassium* – temporarily trapped in clay particles. *Phosphorus* – held on positively-charged zones on the surface of clays.	Not well defined

Table 4.1   *(Cont).*

Restricting factor	Effects on specific nutrients	Crop responses to supplements from manures or fertilizers
Physical:		
Pans (compacted layers)	*All nutrients*, but immobile nutrients such as *phosphorus* are affected most seriously because roots cannot penetrate the pan and phosphates in or below the pan diffuse upwards very slowly.	Some response to water-soluble fertilizers because of increased concentration of nutrients above the pan. Response is usually restricted, e.g. by waterlogging or limited root development. Top dressing of nitrogenous fertilizer gives some benefit by stimulating a root system in the surface soil but the pan *must* be broken for long-term benefits.
Large clods or natural soil- structure units which roots cannot penetrate	*All nutrients*, but immobile nutrients such as *phosphorus* are affected most seriously.	Nitrogenous fertilizers give limited short-term benefit by encouraging root penetration. Long-term benefits from large, repeated applications of farmyard manure.
Biological:		
Competition from living soil organisms (weeds, soil animals, micro-organisms)	*All nutrients*, but especially *nitrogen*. Soil micro-organisms compete with plants for available nitrogen, especially if fresh organic matter with a high carbon/ nitrogen ratio such as straw has been recently incorporated.	Good response.
Lack of earthworms	*All nutrients*. Earthworms play a vital part in improving soil structure and hence the ability of roots to seek out nutrients.	Any addition of manures will help to maintain or increase the earthworm population.
Immobilization in organic matter	*Nitrogen, phosphorus and sulphur* mainly affected by seclusion in poor quality acidic organic matter such as hill peat or mor humus in podzols and in other peats which have not been drained.	Vigorous response to liming and, in some peats, to drainage. After these operations crops will respond better to fertilizers or manures.
Pests and diseases	*All nutrients* but *nitrogen* deficiency symptoms are the most common result. Uptake is restricted because of effects on the efficiency of the plant. The worst effects are from pests or diseases which attack roots (e.g. nematodes) or stems at ground level.	Limited responses, in the case of root nematodes or fungi, to nitrogenous top dressings but these are no substitute for dealing with the pest or disease.

Table 4.1 (Cont).

Restricting factor	Effects on specific nutrients	Crop responses to supplements from manures or fertilizers
Climatic:		
Leaching resulting from excess rainfall	*All nutrients* are leached to some extent, especially in areas of high rainfall with freely-drained acidic soils. Least affected is *phosphorus*. Most affected is *nitrate-nitrogen* because the soil has no mechanism for retaining it. *Sulphate-sulphur* is also readily leached.	Good responses, but added nutrients, especially if water-soluble, are also subject to leaching.
Waterlogging, resulting from poor drainage or from excess rainfall	*Nitrogen* – through denitrification, lack of mineralization, poor nitrogen fixation. The chemical availability of some elements (e.g. *phosphorus, manganese*) is actually increased by waterlogging but this is usually counteracted by the restriction of root systems.	Little response unless waterlogging ceases. Top dressing of nitrogenous fertilizer gives some benefit except in prolonged surface waterlogging.
Drought	*All nutrients.*	Little response unless the drought breaks or irrigation can be used.
Low soil temperature	Mainly *nitrogen, phosphorus and sulphur*. Release of these nutrients from organic matter by mineralization is most rapid in moist, warm summers. Other nutrient elements are released only slowly from rock minerals at low temperatures. Root systems do not elongate and ramify and thus uptake of all nutrients is restricted.	Little response unless soil becomes warmer.

Processes affecting availability

Weathering of rock minerals

Weathering brings about the decomposition and transformation of rock minerals under the influence of rain, frost, sun, wind, plants and animals. During weathering many complex minerals, especially silicates, are converted to more soluble forms. Particles are reduced in size, giving increased surface areas for plant roots to contact. Some silicates are converted to clays which are capable of retaining cations, in exchangeable forms, available to the plant.

Weathering is thus a slow process which releases available nutrients from rock minerals and may also increase the capacity of the soil for retaining nutrients in available forms.

Cation exchange Both clays and humus are negatively charged and therefore attract cations, which are positively charged, including several important nutrient ions (K^+, Mg^{2+}, Ca^{2+}, NH_4^+ and trace elements). The amount of cations which can be adsorbed by a clay or a type of humus (measured in milli-equivalents per 100 g) is described as the *cation-exchange capacity*. This varies considerably from one type of clay to another and from one type of humus to another. Well-decomposed humus has the highest cation-exchange capacity, two to three times as great as the 'best' clay (montmorillonite). Other types of clay, more common in the British soils (illites, kaolinite) have much lower cation-exchange capacities.

Thus the cation-exchange capacity of British agricultural soils depends largely on the humus content in light soils and on clay content in heavier soils.

Table 4.2 gives some examples of the cation-exchange capacities of soils in the British Isles.

Table 4.2 Cation-exchange capacities of soils.

Soil	Cation-exchange capacity in milli-equivalents/100 g soil
Coarse sand	2–4
Coarse sandy loam	4–7
Sandy loam	6–18
Loam	10–22
Silty loam	13–28
Clay loam	16–40
Clay	25–50
Well-decomposed low-moor peat	180–220
Mor humus	20–50
High-moor peat	30–80

Source: Data from several sources.

Nutrients are held on the cation-exchange complex sufficiently strongly to resist leaching but are available for uptake by the plant either directly or by transfer to the soil solution by exchanging with other ions in the solution.

The process of *cation exchange* is involved in several aspects of nutrient availability. Cation exchange is very rapid and follows the normal laws of chemical reaction. A divalent cation such as calcium (Ca^{2+}) will exchange with *two* monovalent cations such as hydrogen (H^+) thus:

$$Ca\,(clay) + 2H^+ = {H \atop H}(clay) + Ca^{2+}$$

This equation represents the replacement of calcium adsorbed on the cation-exchange complex by hydrogen ions from the soil solution. The Ca^{2+} released into the soil can then be leached, taken up by the plant or readsorbed on the cation-exchange complex.

Leaching, soil acidification and antagonistic effects among nutrient cations are all dependent upon the cation-exchange capacities of soils and the process of cation exchange.

Leaching   Leaching is the removal in solution or suspension of substances from the soil to drains and watercourses. Leaching of cations is greatest in areas of high rainfall, low temperature and from soils with a low cation-exchange capacity (gravels, sands, sandy loams with small humus content, see Table 4.2). In soils of very low cation-exchange capacity indigenous cations, cations from water-soluble fertilizers (NH_4^+, K^+) or from the water-soluble fractions of manures are easily leached. In most other British soils such leaching is slight and in high-humus clays very little removal of cations will occur, provided that the soil is not allowed to become strongly acidic.

Some nutrient *anions*, particularly nitrate and to a lesser extent sulphate, are very susceptible to leaching. The soil has no mechanism for retention of nitrate (NO_3^-) and unless taken up by the plant it will, in the event of heavy rain, move rapidly downward through the soil. This is particularly important when considering the choice and time of application of nitrogenous fertilizers.

The rate of leaching is increased by the application of some water-soluble fertilizers; directly because of their solubility and indirectly because of their acidifying effects.

Leaching will inevitably result in soil acidification except in strongly calcareous soils such as those derived from chalk.

Acidification   Soil acidification is caused by carbonic acid received in rainfall and mineral and organic acids produced during microbial decomposition of organic matter or mineral weathering. It is enhanced by atmospheric pollutants such as nitric acid and sulphurous and sulphuric acids. The use of nitrogenous fertilizers containing ammonium ions (NH_4^+) or urea ($CO(NH_2)_2$) also causes considerable acidification because hydrogen ions are produced during nitrification – conversion to nitrates.

Acidification is most rapid in areas of high rainfall in soils of low cation-exchange capacity. Hydrogen ions dominate the soil solution and, by cation exchange, progressively dominate the cation-exchange complex along with aluminium ions in very acid soils. As a result basic cations (Ca^{++}, Mg^{++}, K^+) are displaced into the soil solution and are leached. Consequently the soil pH falls.

The effects of soil acidification on nutrient availability are very great. In addition to intensified leaching, mineralization of organically bound nutrients is retarded and phosphorus is fixed in forms unavailable to the plant. Also manganese and aluminium become more soluble and become toxic to the plant.

Antagonism   If one or more elements are available in excess this can adversely affect both the uptake and the efficiency of other nutrients. Such *antagonism* usually occurs among cations and may be at least partly explained in terms of cation-exchange reactions. One of the

Figure 4.3   *Effect of increasing potassium fertilizer rate on the magnesium content of potato leaves. Unpublished data from a field experiment, Edinburgh School of Agriculture.*

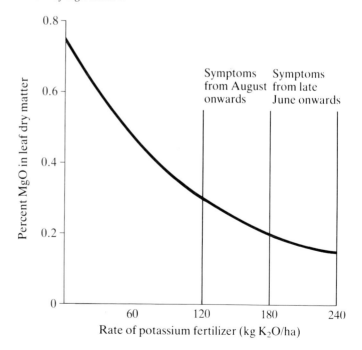

best known examples of antagonism is that of potassium for magnesium. Crops such as potatoes and tomatoes are known to require much potassium. However, large amounts of potassium fertilizer may induce magnesium deficiency in the crop. This may be diagnosed from leaf symptoms – green leaf margins, yellow or brown areas between major veins which also remain green.

There are two reasons for this. Within the soil, potassium and magnesium will compete for sites on the cation-exchange complex. An excess of potassium will cause leaching of magnesium and there will then be less available magnesium in the soil. Also, within the actively growing plant the *total* concentration of cations is fairly constant. The bulk of the cations present consists of potassium, magnesium, calcium and sodium, with potassium dominant in most plant leaves. Potassium is much more readily taken up by the crop than the other elements and is very mobile within the plant. The resultant luxury uptake of potassium will inevitably depress the proportion of magnesium, calcium and sodium in the leaves and it is the less mobile magnesium that becomes deficient. Figure 4.3 gives an example of potassium/magnesium antagonism in potatoes. As the potassium supply to the plant was supplemented by increasing the rate of potassium fertilizer, the concentration of magnesium in the leaf declined. At levels lower than 0.30 per cent MgO in the dry matter the leaves showed magnesium deficiency symptoms with large yellow or brown areas between the major veins.

Other examples of antagonism include calcium-induced potassium deficiency in plants grown on calcareous soils and deficiencies of calcium, magnesium or potassium caused by an excess of hydrogen and aluminium ions in acid soils.

There are even some soils so naturally rich in magnesium (ultrabasic soils) that calcium and potassium deficiency may be induced. Apart from these rare cases and the much more common effects of excess calcium (calcareous soils), almost all deficiencies induced through antagonism are caused by excessive use of lime, neglect of liming, or misuse of fertilizers.

Synergism   When two nutrient elements each reinforce the influence of the other on plant growth they are said to be *synergetic*. Figure 4.4 illustrates a synergetic interaction in which a nitrogenous fertilizer, applied alone to a potato crop growing on a phosphorus-deficient soil, had virtually no effect on yield. When phosphorus fertilizers were also applied, the crop was able to use the available nitrogen,

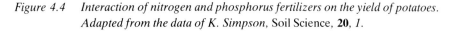

Figure 4.4   *Interaction of nitrogen and phosphorus fertilizers on the yield of potatoes. Adapted from the data of K. Simpson,* Soil Science, **20**, *1.*

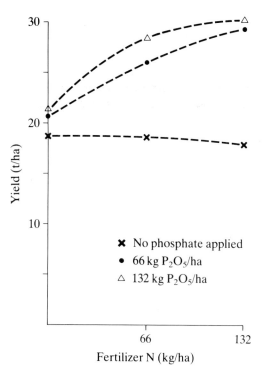

resulting in an appreciable yield increase. Typically, synergetic effects do not improve further once a certain nutrient supply is reached. Thus in Fig. 4.4, doubling the rate of phosphate fertilizer improved the effect of the nitrogenous fertilizer on yield only very slightly.

Figure 4.4 represents the simplest form of synergism. Much more complex interactions occur involving three or more elements. For example, a shortage of potassium can, in some crops, cause iron deficiency in plant leaves. This is probably the result of the failure of enzyme systems involving not only iron and potassium but also phosphorus. Because of this breakdown, iron can enter the plant but gets no further than the roots or stems, thus becoming deficient in the leaves.

Synergism between plants and micro-organisms The most important synergetic effect in crop production involves the harmonious and beneficial living together of plants and bacteria, *Rhizobia*, which inhabit nodules on the roots. This type of synergism is quite different from those which occur between nutrient elements as it involves living organisms. The plants involved are mostly leguminous: clover, peas, beans, lucerne, lupins. Some non-legumes have similar root nodules but the species are not important in agriculture.

In return for some sustenance supplied by the plant the *Rhizobia* take nitrogen from the atmosphere of the soil and convert it into forms available to the plant. The quantities of nitrogen converted can be great. In a vigorous grass/clover sward as much as 150–200 kg/ha of nitrogen can be involved annually. Leguminous arable crops take so much nitrogen from the atmosphere in this way that they seldom require any further nitrogen from manures or fertilizers.

The available nitrogen produced in this way can frequently carry over to reduce the requirement of the following crops for nitrogenous fertilizer. The legume/*Rhizobia* synergism is very complex and involves several nutrient elements, including calcium, phosphorus and the trace element cobalt which is not essential for other species but is essential to legumes because the *Rhizobia* require it in order to survive and multiply.

Synergism between plants and nutrient elements Some plant species are capable of absorbing more nutrients from a soil than others. In some cases this can be explained simply in terms of the vigour and intensity of the root system which enables some species to tap much greater volumes of soil than others. Surface- rooting species such as potatoes and brassica crops are able to colonize much smaller volumes of soil than the deep-rooting lucerne (alfalfa) and cereal crops. Also the intimately ramified root systems of long-established grassland ensure a much more satisfactory assimilation of available nutrients than those of newly sown arable row-crops.

Other differences in uptake of nutrients between species grown on the same soil are the result of substances exuded from the roots (root exudates) into the soil which dissolve nutrients and render them available to the plant. These exudates are usually organic and acidic. An example is found in the ability of the rape crop (*Brassica napus*) to absorb considerably more phosphorus from a given soil, especially if available phosphorus is low, than other species.

Fixation The term 'fixation' is used in this book to indicate the rendering unavailable of nutrients to the plant, either temporarily or permanently, by reaction with soil constituents or living organisms.

It includes, therefore, such diverse actions as temporary immobilization in the bodies of soil animals, fungi and bacteria, the conversion of ions to chemical forms of low solubility and the temporary trapping of ions between the plate-like layers of clay minerals.

Fixation by organisms Temporary fixation in the bodies of soil organisms may affect all nutrients. They are released again when the organism dies and decays. It has been estimated that some 10–30 per cent of fertilizer nutrients are temporarily 'fixed' in this way. Even more fertilizer nitrogen may be immobilized for some months if organic materials such as straw have been recently incorporated in the soil.

Potassium fixation Another example of temporary fixation is the trapping of potassium ions between the plate-like units of some clays, mainly illitic. The fixation tends to occur as the soil dries in the summer and some of the potassium is released after the soil becomes wet in the autumn followed by freezing and thawing during the winter.

Phosphorus fixation Much more serious and permanent fixation of phosphorus occurs in strongly acid soils (pH <5.5) and in calcareous soils (pH >7.0). The processes involved are very complex and have attracted vast amounts of research but the basic position can be summarized quite briefly.

Water-soluble phosphate added to soil in fertilizers, such as superphosphate, mono- and di-ammonium phosphates, is rendered insoluble in the course of days, sometimes hours. It does, however, remain in reasonably available forms for some weeks or months with its availability gradually declining.

The rate of fixation depends very much on soil conditions and particularly on pH. The most serious fixation occurs in very acidic soils (pH <5.0) in which there are large quantities of soluble iron and aluminium hydroxides. These combine with phosphate ions to give iron and aluminium phosphates which have very low solubility, particularly after they have 'aged' in the soil and have become crystalline. This is a gross over-simplification of phosphate fixation in acid soil but does give the essence of it. This type of fixation declines as the pH increases because of the smaller amounts of iron and aluminium hydroxides in solution.

In the middle range of soil pH, 5.5–6.5, fixation is at a minimum but as the pH rises above 6.5 under the influence of excessive liming or naturally calcareous parent material a different type of fixation occurs. This results from the reaction of soluble phosphates with calcium to give insoluble phosphates such as tricalcium phosphate $(Ca_3(PO_4)_2)$. This type of fixation is most serious in soil derived from chalk or soft limestones such as oolite.

Phosphates may also be fixed by silicate clay minerals such as illite or montmorillonite. Some of this phosphate as well as that fixed directly by iron and aluminium can again become available to the plant by a process known as anion exchange by which an anion such as hydroxyl (OH^-) or fluoride (F^-) replaces the phosphate on the clay and thus releases it.

Fixation of trace elements   The chemical forms of trace elements in soil are very strongly influenced by pH and by the availability of oxygen. In general the higher the soil pH and the drier the soil, the less available are the trace elements. There is one exception, molybdenum, which is converted into less available forms at low soil pH, i.e. in acidic soil. Copper, manganese, iron, boron, zinc and cobalt all become less available at high pH in aerobic soil conditions, because they are converted to less soluble forms. The conversion may be purely chemical but, certainly in the cases of iron and manganese, bacterial oxidation can occur. Certain bacteria oxidize ferrous iron (Fe^{2+}) to the ferric form (Fe^{3+}), which is less soluble. The bacteria involved operate best at high soil pH. Similarly manganous manganese (Mn^{2+}) may be oxidized by specific bacteria to the less soluble manganic form (Mn^{3+})

Mineralization

Mineralization is the conversion of nutrients contained in complex forms in soil organic matter to simple ionic forms available to the plant. The main elements involved are nitrogen, phosphorus and sulphur. It is important that the release of nutrients occurs through the action of soil micro-organisms and will *not* occur if their action is restricted by high carbon/nitrogen ratios, by soil acidity or by poor aeration.

Mineralization is most rapid in moist soils of moderate–high pH (5.5–7.5) immediately after the incorporation of green manures or farmyard manure already partially rotted. Release of nitrogen, phosphorus and sulphur from the easily decomposable low C/N ratio organic material is rapid.

Much slower mineralization of more resistant organic matter reserves varies from season to season. It is greatest in warm, moist summers. Obviously, in a season when mineralization is vigorous, less nitrogen, phosphorus and sulphur need to be added in fertilizers but unfortunately the *amount* likely to be released is very difficult to predict.

Nutrient balance

The term 'nutrient balance' has been sadly overused. It conveys the impression of knife-edged precision in the requirements of plants for a balanced supply of nutrients. In fact, most plant species can grow efficiently under a wide range of ratios and quantities of nutrients available to them. In this range the synergetic interactions of essential elements proceed satisfactorily and antagonistic interactions do not reduce the concentration of any element in the plant to a critical deficiency level. It is generally only gross excesses of particular elements which vitiate the synergetic actions and bring about nutrient deficiencies through antagonism, causing serious crop problems. Such excesses can be caused by natural soil properties such as calcareous parent material. Unfortunately, however, the most common sources of gross excesses are ill-planned liming programmes and fertilizer treatments concentrated largely on the excessive use of nitrogen, phosphorus and potassium and neglecting other essential elements.

Chapter 5 The supply of nutrients to plants

The seed as a source of nutrients

In the very early stages of growth the plant draws much of its nutrient supply from its own seed. Evidence of this is that oat plants grown in soil acutely deficient in manganese show no symptoms on the first two leaves produced. They have drawn upon the limited supply of manganese in the parent seed. This source is, however, rapidly depleted even in crops such as potatoes, propagated from 'seed' tubers which contain more nutrients than most true seeds. The plant with its rapidly developing root system must then depend upon what it can obtain from the soil or from supplements such as lime, manures and fertilizers.

The soil as a source of nutrients

Even the poorest soils can supply sufficient essential elements to support some kind of vegetation, as for example the beautiful but sparse flora growing in the crevices of the Burren limestone area of western Ireland, which is useless for agriculture. The poorest *agricultural* soils in the British Isles are, perhaps, the very acid peats of north-west Scotland which can supply nutrition only for communities of heather, cotton grass and sphagnum moss. The annual dry matter production of these soils is only 0.5–1 t/ha and they are capable of supporting only 1 sheep per 1–2 hectares. In these soils nutrient supplies are very poor indeed and excess water brings about conditions in which roots are asphyxiated and cannot take up the small amounts of nutrients that are available. In the very acidic conditions the bacteria which mineralize organic nitrogen, phosphorus and sulphur compounds to forms which the plant can use are not able to operate because they need both oxygen and a relatively high pH. Thus, although the peat may contain appreciable quantities of *total* nitrogen it is locked up in complex organic compounds and unavailable to the plant. Because of the depth of the peat and the inability of the plant roots to reach underlying mineral soil or rocks, its total trace element content and even that of major elements such as phosphorus and potassium is very low.

These soils are very difficult to use for agriculture and attempts to improve them by drainage, liming and fertilizer use are usually uneconomic.

Another extreme type of soil in terms of limited nutrient supply is the very dry freely-drained soil with chalk or limestone parent materials which supports poor grass-heath in southern England. The vegetation communities are dominated by grasses such as brome, sheep's fescue and false oat grass and low growing herbs such as marjoram, thyme and calamint.

These soils are little more productive than the acid peats, but for entirely different reasons. Availability of nutrients is restricted here by drought, by the fixation of phosphorus and the reduction in availability of potassium and trace elements resulting from the high soil pH. The supply of available nitrogen is also restricted by the small turnover of organic nitrogen, in this case because it is too dry in summer for intensive bacterial activity.

These two extreme cases illustrate many of the factors which affect the ability of soils to supply nutrients to the plant: excessive wetness or dryness, too high or too low a pH value, the presence of available forms of the nutrient. Between the two extremes lie the whole range of agricultural soils, each capable of supplying nutrients to produce a limited crop yield year after year without the use of fertilizers or manures. This basic sustained yield potential varies a good deal from soil to soil, depending on the rate at which reserves of various nutrients are depleted, and is commonly much too low to be economic without resort to the use of lime, fertilizers and manures.

Areas of calcareous soils where no fertilizer or manures have been used for more than 100 years, as for example the classical plots at Rothamsted (Broadbalk) sustain wheat yields of only 1.5–2 t/ha despite the use of modern varieties and methods of husbandry.

Vast areas of non-calcareous soils rely on regular liming to maintain their capacity to supply nutrients. Before the great liming campaigns of 1938–65 many of these soils were very acid and the resultant deficiencies of available phosphorus or potassium and the toxicity of aluminium and manganese caused widespread crop failures. If liming were discontinued now these soils would revert from their present state over a period of 10–20 years to infertility and, as in the 1930s, barley and sugar beet crops with their high lime requirement could be grown only on naturally calcareous soils. Only small crops of acid-tolerant species such as oats or rye would be sustainable on soils which do not have calcareous parent material.

Nutrient reserves in the soil   The basic capacity of a soil to supply a nutrient over a period of many years depends upon the presence of a large *reserve* of that nutrient in the soil or its parent material and the *rate* at which it is converted to forms available to the plant. The reserves come from two sources: the mineral parent material of the soil and the residues from previous applications of lime, fertilizers, manures or crop residues.

Nitrogen   Nitrogen is a special case. There are virtually no reserves of nitrogen from the parent materials of mineral soils but it is contained in the organic matter residues of previous crops, manures, and the decaying remains of dead soil organisms such as fungi and bacteria.

Even the poorest soils will supply some nitrogen for the crop by virtue of the small amounts of nitrate, 3–10 kg/ha per year, which the soil receives in rainfall. This has been formed during electrical storms and from industrial pollutants. There are also nitrogen-fixing bacteria in the soil, both free living and in the root nodules of legumes, which convert atmospheric nitrogen to available forms. Other bacteria will release small amounts of nitrogen, even in very adverse soil conditions, by mineralization of complex nitrogenous compounds in the soil organic matter.

Among the lowest nitrogen-status soils are the young dune sands in which vegetation is only beginning to establish itself. As a result little organic matter is present and bacterial action is low. For quite different reasons deep acidic hill peats also supply little available nitrogen. Such soils may be expected to supply only 3–10 kg nitrogen/ha per year to the plants growing on them.

In contrast, a neutral or slightly acid soil, pH 6.0–7.0, which has been under pasture rich in clover for five or six years can release as much as 200 kg/ha of nitrogen to a crop grown in the year after ploughing out. The reasons for this are the build-up of organic matter in the soil under a grass/clover ley, the release of available nitrogen from that organic matter by bacteria and the further supply of nitrogen from the *Rhizobia* bacteria in the clover root nodules. The amount of nitrogen released can be in excess of crop requirements and may lead to lodging of cereals.

Most soils lie between these extremes and may be expected to contribute some 50–150 kg/ha of nitrogen each year to the crop. The ability of the soil to supply nitrogen will depend on:

● Previous manuring.
● Rainfall and soil texture which affect leaching of nitrates.
● Previous cropping, especially recent legume crops such as peas, beans, clover.
● The amount of organic matter in the soil.
● The soil pH and temperature which affect bacterial activity.
● The ability of the crop to produce an efficient root system.

Table 5.1   Amount of nitrogen supplied to crops by soil without supplementary fertilizer nitrogen.

Conditions	Nitrogen supplied (kg/ha)
Continuous cereal cropping for ten years, straw removed or burnt	20–40
Continuous cereals, straw returned to soil regularly for ten years	40–80
Poor permanent pasture, soil pH 5.0, no clover, ploughed in previous autumn	Nil–20
Previous crop cereals, followed by leguminous green manure crop	60–90
Moderate dressing of good farmyard manure (25 t/ha) or slurry (20 000 litres/ha) applied for the crop	80–120
Large dressing of good farmyard manure (40 t/ha) or slurry (30 000 litres/ha) applied for the crop	120–150
Previous crop sugar beet, tops removed	40–60
Previous crop sugar beet, tops ploughed in	80–100
Previous crop peas or beans, good yield	80–100
Very rich clovery long-term ley, ploughed in previous autumn	150–200
Long-term ley with little clover but ploughed in previous autumn after several years of high nitrogen input	150–200
Fen peat rich in organic matter	130–160

Source: Data from several sources.

As a rough guide Table 5.1 gives examples of the nitrogen-supplying power of soils in a range of circumstances.

For the great majority of soils, arable crops and seasons the supply of nitrogen from the soil is exceeded by the amount required by the crop, especially during the period of rapid vegetative growth. This imposes the need for additional nitrogen in the form of manures and fertilizers. Assessing *how much* extra nitrogen will be required is a major problem and making an accurate estimate will probably contribute more to high crop yields than any other single factor under the farmer's control.

Phosphorus

Many of the soils of the British Isles are fundamentally deficient in phosphorus. As a result they are largely dependent for their phosphorus-supplying capacity on residues from past applications of fertilizers or manures. In many hill and upland areas and some lowland areas with a long history of soil acidity the amount of phosphorus available is insufficient to support even modest crop yields and phosphorus fertilizers become essential. The same would be true of many areas of calcareous soil. Such soils under the best cropping conditions will supply only 4–8 kg/ha P_2O_5 each year.

In contrast to nitrogen, the loss of phosphorus from most soils by leaching is negligible, less than 1 kg P_2O_5/ha per year although some losses occur from peat and from very sandy soils. Because of this, together with the relatively poor uptake of fertilizer phosphorus by crops, there is an inevitable build-up of *residual* phosphorus in the soil. In areas of high (>6.5) or low (<5.5) soil pH values much of the residual phosphorus is 'fixed' in unavailable forms and crops can obtain only a very small proportion of it each year. In some arable areas, however, where good liming practice has maintained soils in the middle pH range (6.0–6.5), much residual phosphorus is available to crops, as indicated by moderate, high or very high soil analysis results for available phosphorus. These soils have a legacy of high, often excessive use of phosphate fertilizers from the period 1930–70 when they were extremely cheap. Assuming an average annual input of 100 kg/ha of fertilizer P_2O_5 over this 40-year period, an uptake by the first crop of 20 per cent of the amounts applied and uptakes of about 1 per cent by each subsequent crop, the reserve of phosphorus in the soil would have been increased by some 2 400 kg/ha – more than half that applied during the 40-year period.

Soils with high or very high available phosphorus by soil analysis are capable of supplying more than 50 kg/ha of P_2O_5 per year and could undoubtedly support crops for some years without the use of phosphate fertilizers. This, however, is not to be recommended, and maintenance applications should be made to replace the phosphorus removed by the crop.

There are strong indications from soil analysis results that the phosphorus-supplying power of soils of the British Isles has greatly improved over the last fifty years, largely as a result of judicious liming to control pH and the build-up of phosphate residues. Many of our soils, however, are still deficient in phosphorus mainly because of fixation in the form of unavailable iron and aluminium phosphates (acid soils) and insoluble tri-calcium phosphate (alkaline soils). The soils mainly affected by acid fixation are in marginal or hill areas, particularly in the wetter parts of the north-west. It is minimized by liming the soil to the pH range 5.5–6.5. There is no such simple palliative for the type of fixation which occurs mainly in naturally calcareous soils. The amount of calcium carbonate in these soils is so great, commonly more than 80 per cent of the whole soil, that there is no practical way of reducing the pH and thus avoiding phosphorus fixation.

As with nitrogen, many other factors besides the chemical forms affect the amount of phosphorus that can be drawn from a particular soil by the crop. In contrast to the extremely mobile potassium and nitrate ions, phosphate is present in the soil solution in very low concentrations. It moves in the soil only very slowly. Thus, the plant roots must seek out soil phosphorus and the ability of the crop species to produce an extensive and well ramified root system early in the season becomes a dominant factor. This is typified by vigorous permanent grassland or long-term leys with intensive root systems which can remove phosphorus from the soil very efficiently. So much so that, in a long-term field experiment on a ley in south-east Scotland grass yields were maintained at a high level for seventeen years (12–13 t/ha of dry matter each year) by using only nitrogen and potassium fertilizers. Plots which also received phosphorus fertilizer yielded no more than those which did not. The original soil, although it was classified as only 'moderate–low' in available phosphorus, was able to supply sufficient phosphorus (40 kg/ha P_2O_5 per year on average) to support intensive grass production for this period from its original phosphorus content. This is seldom the case with arable crops with their less intensive root systems, although some crops extract phosphorus from the soil more efficiently than others.

Other factors affect the ability of a crop to withdraw phosphorus from the soil. In seasons of high soil temperature and adequate soil water large amounts of organically-bound phosphorus are mineralized to available forms. Good soil structure and available soil water encourage root penetration and stimulate phosphorus uptake, as does an adequate supply of available nitrogen for the crop.

Figure 5.1   *Effect of nitrogenous fertilizer and irrigation on the P_2O_5 uptake of potatoes grown without phosphorus fertilizer. Derived from the data of K. Simpson, J. Sci. Food and Agriculture, **13**, 236.*

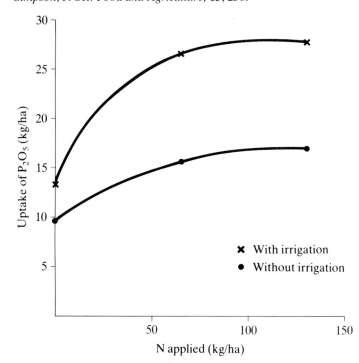

Because of the many factors involved it is impossible to predict from soil analysis the precise amounts of phosphorus that a particular crop will obtain from a particular soil in a particular season. This is graphically illustrated in Fig. 5.1 showing the results of a field experiment on potatoes in which various rates of fertilizer nitrogen and phosphorus were used with and without irrigation. The amount of phosphorus taken up by the crop, especially when no fertilizer phosphorus was used, was very strongly influenced by the amounts of both water and nitrogenous fertilizer available to the crop. Figure 5.1 shows that the actual amount of P_2O_5 per hectare obtained by the potato crop from this soil which had a moderate–low P_2O_5 value by soil analysis (ADAS Index 1) varied from less than 10 kg/ha to 28 kg/ha– an almost threefold variation at one site in one season. Indeed on some areas (not shown in Fig. 5.1) which had no fertilizer phosphorus, the uptake was *greater* than on areas which received normal rates of phosphorus fertilizer. In the light of such evidence as this, soil analysis results must be regarded only as a very rough guide to the *comparative* phosphorus-supplying capacity of soils.

Fortunately, the most reliable interpretations can be put upon soils found to be very low or low in available phosphorus. These soils will supply only 4–8 kg/ha of P_2O_5 each year and the inevitable result, unless fertilizer or manure is used, is crop failure or low yield.

Yield responses to phosphorus fertilizers on such soils are quite dramatic. Some examples are shown for swedes in Fig. 2.2 (p. 17). Obviously the economic value of such responses is beyond question. The one exception is the poor response on soil X which was a result of soil acidity causing a near failure of the crop.

Potassium Most of the soils of the British Isles contain appreciable reserves of potassium but much of it is held in complex rock minerals from which it cannot be extracted by the plant. Soils most abundant in available potassium are those derived directly from rocks rich in potassic micas and potash feldspar. Some mica schists in the north of Scotland contain as much as 12 per cent of K_2O. The annual weathering of these minerals is sufficient to ensure that the soils containing them have ample available potassium.

Such soils are not common in other areas but there are large areas of parent materials and soils which contain weatherable reserves of potassium minerals, supplemented by fertilizer residues, sufficient to supply some 40–120 kg/ha of potassium to the crop each year. This is an appreciable proportion of the requirements of crops (Table 3.1, p. 23), especially cereals.

In contrast some mineral soils derived from chalk, limestone, siliceous sandstones or fluvio-glacial sands and gravels have very low reserves of potassium and release only 20 kg or less of potassium per hectare each year. Unfortunately, because of the porous nature of these soils and their low cation-exchange capacity, potassium is easily leached from them. This is particularly so in the chalk and limestone soils because of calcium/potassium antagonism. Some peats, especially those deep enough to prevent access of plant roots to underlying mineral materials, are also deficient in potassium.

Such low-reserve soils, which usually show low or very low values by soil analysis, cannot satisfy the requirements of even the least demanding of cereal crops without the use of manures or fertilizers.

Soil analysis is not a very reliable guide to the amount of potassium a crop will obtain from high-reserve soils. Readily available potassium can be rapidly depleted in such soils by, for example, growing intensive grass with high rates of nitrogen fertilizer and inadequate potassium fertilizer. This can result in an autumn soil analysis giving a very low value but weathering of reserves during winter and spring can release sufficient potassium to supply 40 or more kg K_2O/ha for the next crop. Soil analysis for available potassium must, therefore, be interpreted in the light of the nature of the soil and its history of fertilizer and manure applications.

Apart from sandy, chalk and limestone soils low in organic matter, leaching of potassium is not a major problem. Undoubtedly some leaching does occur in a very wet spring before crops have become established and where large applications of farmyard manure, potassium-rich slurries or fertilizers have been made. Potassium is, however, quite strongly adsorbed on the cation-exchange complex from which plants may draw upon it freely. Thus in humus-rich soils, especially with a high clay content, leaching is usually negligible.

Sulphur

Some sulphur is contained in the soil in the form of inorganic sulphates (SO_4^{2-}) and sulphides (S^{2-}) but a large proportion is present in the form of organic compounds which are residues from previous crops and manures. This sulphur becomes available to the plant only when it is oxidized by bacteria to inorganic sulphates.

These sulphates are subject to leaching, although less easily leached than nitrates because they can be held loosely by the soil on positively-charged sites on clay or other minerals. None the less there is little residual available sulphate in the soil following a wet winter.

In conditions of waterlogging caused by poor drainage, organic sulphur compounds and sulphates may be reduced to sulphides, especially hydrogen sulphide (H_2S) which has a characteristic 'rotten egg' smell and is toxic to plants. This smell is immediately evident in a newly-dug spadeful of shallow peat or in a layer of farmyard manure or straw badly incorporated in waterlogged soil.

There is no doubt that the amounts of sulphur supplied by many British soils would be insufficient for the needs of crops were it not for atmospheric pollution and the organic residues from crops grown more than two decades ago when the sulphur content of fertilizers was high. Now that low-sulphur fertilizers are predominant and pollution has been reduced it is predictable that many soils, particularly in low pollution areas, will become progressively deficient in sulphur. Monitoring of sulphur pollution in the British Isles indicates that it is greatest in areas of high rainfall and areas to the lea of installations such as large coal-fed power stations. Areas of high risk of sulphur deficiency include most of Ireland, south-west England, parts of north-east England and much of the east of Scotland.

Calcium and magnesium

Calcium and magnesium occur widely in rock minerals which, when weathered, release them in forms available to the plant. Calcium and magnesium carbonates ($CaCO_3$, $MgCO_3$) in various proportions are the main constituents of chalk and limestones. Soils derived from these materials will supply more than enough calcium for crops. This is not always true for magnesium because chalk and the many limestones classed as calciferous have very high ratios of calcium to magnesium. This gives rise to calcium/magnesium antagonism and deficiencies of magnesium may occur.

Non-calcareous soils are based on parent materials which contain little or no calcium carbonate or magnesium carbonate. They rely for their natural sources of calcium and magnesium on rock minerals such as calcium feldspars and ferro-magnesian minerals. The availability of this calcium and magnesium is dependent on two conflicting factors – the rate of weathering and the rate of leaching. In the cool, wet climate of northerly and westerly parts of the British Isles leaching tends to dominate, especially in low-humus sandy soils. As a result, all except calcareous soils become acidic and, in the extreme, supplies of calcium and magnesium become inadequate for good crop growth. More commonly other aspects of soil acidity, particularly the release of large amounts of manganese and aluminium, cause crops to fail and liming becomes necessary.

Like potassium, calcium and magnesium are retained in available form by the clay/humus as exchangeable cations.

Because of regular liming, calcium deficiency is seldom now the cause of crop failures, but magnesium deficiency can become serious, especially if no additions are made in fertilizers or as magnesian limestone, containing magnesium and calcium in roughly equal amounts.

The trace elements

Because of the relatively small requirements of crops (e.g. less than 0.1 kg/ha of molybdenum, 0.4–1.0 kg/ha of manganese), most of the soils of the British Isles can supply sufficient trace elements. For example, the *total* amounts of iron and manganese in most acidic mineral soils are sufficient to supply crops for many thousands of years.

It is when soils are mistreated, for example by overliming or by cultivating with heavy implements when too wet, that trace element deficiencies arise. In the case of overliming the trace elements manganese, boron, copper, iron and zinc are altered chemically to forms which are less available to the plant. In the case of poor cultivations, consequent effects on soil structure make root penetration difficult or impossible and the plant is physically prevented from obtaining the less mobile trace elements.

Some types of soil are fundamentally deficient in certain trace elements. One of the best examples of this is the very low *total* copper content of some uncultivated deep peats, strongly leached podzols under heather and some soils developed on limestones. If such soils are reclaimed for agriculture, crops grown on them may develop copper deficiency symptoms. In fact one of the names for copper deficiency is 'reclamation disease'.

The failure of soils to supply sufficient trace elements for the crop is commonly the result of high soil pH values (>6.5) whether through over-liming or through having a naturally calcareous parent material. All trace elements, except molybdenum, are less available to the plant from alkaline or very mildly acid soils than from more acidic soils. Figure 5.2 shows the relationship, over a wide range of soils, between soil pH values and the concentration

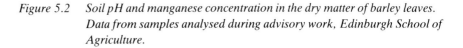

Figure 5.2 *Soil pH and manganese concentration in the dry matter of barley leaves. Data from samples analysed during advisory work, Edinburgh School of Agriculture.*

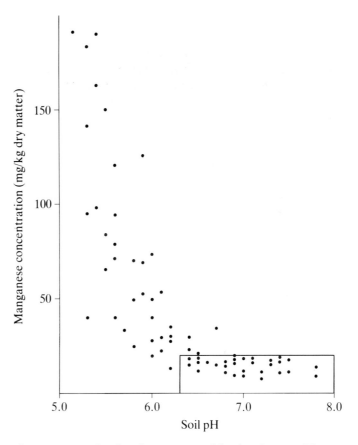

of manganese in the dry matter of barley leaves. The oat crop prefers to grow in soils of pH range 5.5–6.0 and is particularly susceptible to manganese deficiency if the pH is raised above this. Yields can be greatly reduced as a result. Barley is slightly less susceptible and wheat less so again.

The samples represented in Fig. 5.2 were taken in the course of advisory work from crops which were showing, wholly or in parts of the field, visual manganese deficiency symptoms. All samples in the 'box' at the bottom right hand corner of Fig. 5.2 showed typical symptoms – roughly oval dark brown lesions, 1–2 mm in length, apparently distributed at random along the veins. The criticial concentration of manganese in leaf dry matter is approximately 20 mg/kg. Most of the cases occurred on light-textured soils and many were associated with dry periods of weather. In these conditions at high soil pH values manganous ions (Mn^{2+}), which are easily available to the plant, are oxidized to the manganic form (Mn^{3+}) which is insoluble and very much less available to the plant.

Similar diagrams to Fig. 5.2 could be presented for other trace elements and it may be concluded that soil pH has a major influence on their availability.

The atmosphere as a source of nutrients

The atmosphere supplies the water and carbon dioxide essential for photosynthesis by plants. Also, very variable amounts of gaseous nitrogen from the soil atmosphere can be converted to forms available to the plant by bacteria, living freely within the soil and contributing 10–40 kg N/ha each year. Much greater contributions are made by symbiotic *Rhizobia* bacteria living in root nodules of legumes and capable of converting up to 200 kg of elemental nitrogen each year to available forms.

Apart from these major contributions to plant nutrition, the atmosphere supplies variable but often significant amounts of other elements to the soil each year. In islands such as the British Isles there is an influence at least for several kilometres inland of wind-borne sea spray. In extreme cases this can be detrimental because of excess sodium but such cases are limited to exposed areas very near the sea. Further inland useful amounts of sodium, potassium, sulphur, magnesium and trace elements may be deposited.

Other major contributions, particularly of available nitrogen and sulphur, come from pollution. 'Acid rain' has received much well-deserved adverse publicity. The acids involved include sulphuric acid and nitric acid derived from the combustion of fossil fuels. Both are strong mineral acids and very powerful pollutants. On either a country-wide or a continental scale it is difficult to predict exactly where such pollutants will reach the soil. A large coal-fuelled power station can put into the atmosphere several tonnes of sulphur *per day* mainly in the form of sulphur dioxide (SO_2) which dissolves in water to form sulphurous acid (H_2SO_3 easily oxidized to sulphuric acid, H_2SO_4). In simple terms the sulphur dioxide thrown up in one year from one such power station, if it could be properly harnessed, could supply sufficient sulphur for crops on more than 25 000 ha of sulphur-deficient land. At present we can only accept it where it falls and the evidence shows that most of it comes down in high-rainfall areas, leaving low-rainfall areas more susceptible to sulphur deficiency.

The amounts of various nutrients coming into the soil from the atmosphere vary very much from place to place. The highest concentrations are those of nitrogen and sulphur. Chlorine is also found, in the form of chlorides, in large concentrations near the sea. Smaller amounts of phosphorus, potassium and the trace elements are also found. In the vicinity of some industrial plants, such as smelters, the concentration of some trace elements, for example zinc, becomes alarmingly high.

Thus nutrients received by the soil from the atmosphere must be regarded at best as an unpredictable bonus and at worst as major pollutants.

Manures as a source of nutrients

Manures produced on the farm, whether with straw or other bedding materials (farmyard manure) or by diluting raw excreta with water (slurry), contain the full range of nutrients needed by the plant, but not necessarily in desirable proportions. An appreciable amount of these nutrients (20–30 per cent of the nitrogen) is available for the first crop after planting.

The concentration of nutrients in manures is low and applications of 25–40 t/ha are needed to supply some 20–30 per cent of the requirements of a potato crop. Not all the nutrients contained in farmyard manures and slurries are immediately available to the crop. For example, 70–80 per cent of the nitrogen in farmyard manure and 50–70 per cent of slurry nitrogen are in forms which become available only with difficulty. This part of the nutrient content will not be available for the first crop after planting but will increase the reserve of nutrients in the soil.

Fertilizers as a source of nutrients

Fertilizers are made specifically to supply one or more nutrients to supplement what can be derived from the soil. In contrast to manures, most modern compound fertilizers aim to supply only nitrogen, phosphorus and potassium, three of the major elements, but neglect the other major elements (calcium, sulphur, magnesium) and the trace elements. Thus the term 'complete' fertilizers, used by some manufacturers until recently, is very misleading.

The great majority of compound fertilizers (supplying NPK) or 'simple' fertilizers (supplying one nutrient element only) are water-soluble and readily available to the plant. The residues from these fertilizers, not taken up by the first crop, are either leached from the soil or are converted to less available forms by processes such as fixation or conversion to organic forms by soil organisms.

Other fertilizers, mostly phosphates, are deliberately designed to boost the reserve or 'available with difficulty' parts of the total nutrient content of the soil. These fertilizers rely on normal soil weathering processes and become slowly available for crops over several years. They include ground mineral phosphates and basic slag.

At present sulphur, calcium, magnesium and the trace elements are supplied mainly in special fertilizers outside the normal range produced by major fertilizer companies.

In contrast to manures, the NPK contents of fertilizers can be accurately controlled and the amounts of a nutrient added to the soil in a given amount of fertilizer are very precise. If added to a near-ideal medium such as vermiculite, plant nutrition can be closely controlled. Unfortunately there is no ideal soil and the amounts of fertilizer nutrients remaining easily available for a given length of time vary greatly from soil to soil.

Chapter 6 Assessment of availability of nutrients

Because of the many factors involved (see Table 4.1, p. 31), any attempt to assess the availability of a nutrient by a single simple test in the laboratory or in the field is fraught with difficulties. Figure 5.1 (p. 50) illustrates that the amount of phosphorus taken up from a soil by potatoes may be varied almost threefold at a single site in the same season, simply by adjusting the amounts of water and fertilizer nitrogen available to the crop.

Thus soil analysis, plant analysis and other more expensive methods of putting a single value on the amount of an 'available nutrient' in a soil must be regarded only as a guide to the ability of that soil to supply the nutrient in comparison to other soils.

Soil sampling and analysis

The aim of soil analysis is to classify soils into broad groups in terms of the amounts of 'available nutrients' removed from the soil by a particular extractant. Many of the methods used are empirical. The analyst attempts, by using one of the hundreds of methods developed during this century, to extract from the soil a fraction of the total amount of a nutrient which is closely correlated to that actually used by the plant in the field.

Figure 6.1 shows the amounts of P_2O_5 extracted from forty-six different soils by ammonium acetate/acetic acid buffer solution in the laboratory plotted against the actual amounts of P_2O_5 in kg/ha taken up by swedes grown on control plots (no fertilizer used) in field experiments. The relationship, although positive, is not impressive. In fact better correlations between soil analysis and field performance cannot be expected, particularly as the series of experiments in Fig. 6.1 were made in ten separate seasons.

Thus, the selection of suitable methods of analysis is difficult enough because of the wide range of substances in the soil and the variations in field performance. But, more important, *soil analysis is useless unless performed on a good sample of the field or area in question.*

Soil sampling   Soil sampling is a very skilled job which seldom receives adequate attention. Many of the samples analysed each year give false results because of inadequate sampling by unskilled, untrained and, occasionally, unscrupulous samplers.

Figure 6.1   *Phosphorus uptake of swede turnip crops at 46 sites in relation to soil analysis results. Unpublished data accumulated during investigations associated with paper by K. Simpson, J. Sci. Food and Agriculture, 7, 745.*

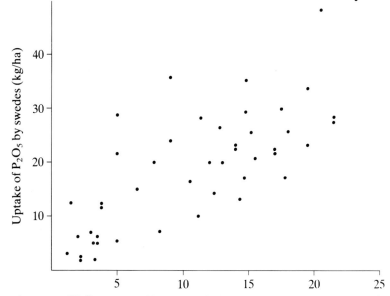

Amount of P_2O_5 extracted by ammonium acetate/acetic acid (mg/kg soil)

Soil is so variable from one place to another within a field that it is very difficult to get an adequate sample from an area.

Whether the samples are being taken for a routine check on lime and major element status or for a more detailed trace element analysis of 'good' and 'poor' areas in problem cases, it is necessary to check that the sampling area is reasonably uniform. First a check should be made of the cropping, liming, manuring and fertilizer history of the field. This is particularly important if several fields have recently been made into one. An exploratory augering of the field should be made to identify variations in soil type.

It is important to avoid sampling an area within two years of liming or four months of fertilizer applications. One small particle of lime or fertilizer, inadvertently included in a sample, will create a grossly false analytical result. Except for comparative samples, taken in crop failure cases, sampling is best done in autumn or early winter, avoiding any unrepresentative parts of the field – headlands, around food and water troughs.

Sampling should be done with an auger with a 20–25 cm-long bit 20–30 mm in diameter. Alternatively a corer of the same dimensions can be used but this is not recommended for stony soils. If sampling dry light-textured soils a narrow-bladed fern trowel should be used because the soil would fall away from an auger or corer. In this case a small hole should be dug to plough depth by trowel and a thin slice of soil taken down the side of it.

Figure 6.2   A pattern for soil sampling.

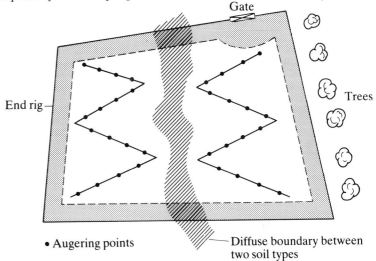

The actual sampling of the selected area is best done on a systematic basis, taking 16–20 auger corings 20–25 cm deep and combining them in one container. Ideally the area from which a single soil sample is taken should not exceed 1 hectare but this ideal is seldom achieved. In routine sampling, one or two samples per field are all that can be realistically achieved. Figure 6.2 illustrates a simple sampling plan for a field in which two very different soil types are found with a diffuse boundary between them. The zigzag path followed by the sampler, taking four or five corings on each traverse, gives a much more representative sample than following a diagonal across the area.

Strong washable linen bags with built-in string ties are ideal containers for the samples. Alternatively, strong waterproof-lined paper bags may be used. Labelling should be done with waterproof inks. It is critically important, if the samples are to be analysed for trace elements, to avoid using metal or other containers that might contaminate the sample.

Soil analysis

No attempt will be made here to describe the analytical methods in use. Modern methods are based on sound theory and are usually automated to ensure a rapid turnover of samples.

The results are intended as a general aid to assessing lime, manure and fertilizer requirements of crops. Routine analysis generally involves the determination of pH, lime requirement, available phosphorus and potassium. Organic matter content, available calcium, magnesium and sulphur may also be determined. Determination of the total quantity of a nutrient in the soil is seldom of any value. It is the proportion of that total existing in forms which the plant can take up that is important. Trace-element analysis is undertaken less frequently, usually if there is cause to suspect a deficiency or toxicity of a particular element.

The most important aspect of soil analysis is, and always will be, the sampling. However advanced laboratory techniques may become, the key to the value of soil analysis is in the variability of soils in the field and the consequent difficulties in getting a good sample. Because of this, all that can be achieved from soil analysis is a rough guide to the likely supply of an element to the crop using terms such as very high, high, medium, low and very low. It is perhaps pretentious to use even these categories – high, medium and low would be more realistic.

Figure 6.3 shows the interpretation scales for available phosphorus (here converted to P_2O_5) and potassium (expressed as K_2O) used by ADAS, who prefer an index system, and the East of Scotland College of Agriculture, who use terms such as 'high' or 'low'. The method of analysis is the same and the scales cover the full range of soils from the most deficient to highly fertile greenhouse soils, but only those mainly relevant to agricultural soils are shown.

The scales are broadly similar. Yield responses to fertilizer would be expected in all crops on soils with very low or Index 0 status and no yield responses would be expected on high status soils (Index 3–5). The critical range in which many soils fall is that between low and moderate (Index 1–2). In this range yield responses to fertilizer will depend very much upon the crop species, soil type and season.

The practical value of soil analysis in deciding on fertilizer rates for crops is discussed in Chapter 15.

Figure 6.3   *Scales of interpretation of soil analysis. Information supplied by P. Crooks.*

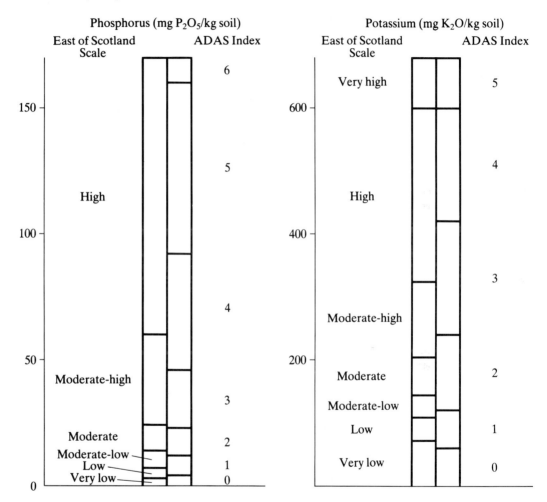

Rapid soil testing   *The hazards of sampling, analysing and interpreting the results are such that any on-the-spot or do-it-yourself methods must be looked upon with grave suspicion. On-the-spot estimation of soil pH may be useful on occasions for the diagnosis of liming problems.*

Rapid testing, done in the field, for available nutrients is likely to be of very little value.

Plant analysis There is no doubt that the analysis of plant dry matter along with the assessment of crop yield would give a reasonably accurate value of the amount of a nutrient taken up by a crop. However, plant tissue varies considerably from part to part (leaf, stem, root, seed, tuber) and also from time to time during the season. Also, by the time plant tissue has been analysed, the crop is well advanced and it is too late to correct any deficiencies indicated by the analysis of the current crop. Plant analysis may give some guidance as to potential deficiency problems in subsequent crops but this is by no means certain because of the many causes of deficiencies (see Table 4.1, p. 31).

Plant analysis is most useful in confirming the cause of visual deficiency symptoms, especially of trace elements in crops and particularly if samples from both healthy plants and those showing symptoms can be compared. For most trace elements, plant analysis is a better guide to the need for supplementation than soil analysis, the only exception being boron.

The same care is necessary in sampling plants as in soils. Plant sampling and analysis should both be undertaken by the specialist with an understanding of the need for sampling the most suitable part of the plant and a knowledge of the effects of stage of growth on composition of the plant. A simple example will suffice: the potassium concentration in young cereal leaves can be four times as great in the first two weeks after emergence as that when ears begin to emerge. Thus time of sampling is critical in the interpretation of results.

Chapter 7 Lime and liming

The pH of the soil is of primary importance for efficient plant growth and nutrient uptake (Table 4.1, p. 31), being a measure of the degree of acidity or alkalinity of the soil: pH 7 is neutral. Figure 7.1 illustrates the range of soil pH values. Nutrient availability, the physical structure of the soil and the activity of micro-organisms responsible for humification are all greatly affected by pH. The most satisfactory combination of these factors is usually found in the pH range 5.5–7.0 and it is within this range

Figure 7.1 *The pH range of some 'natural' and cultivated soils.*

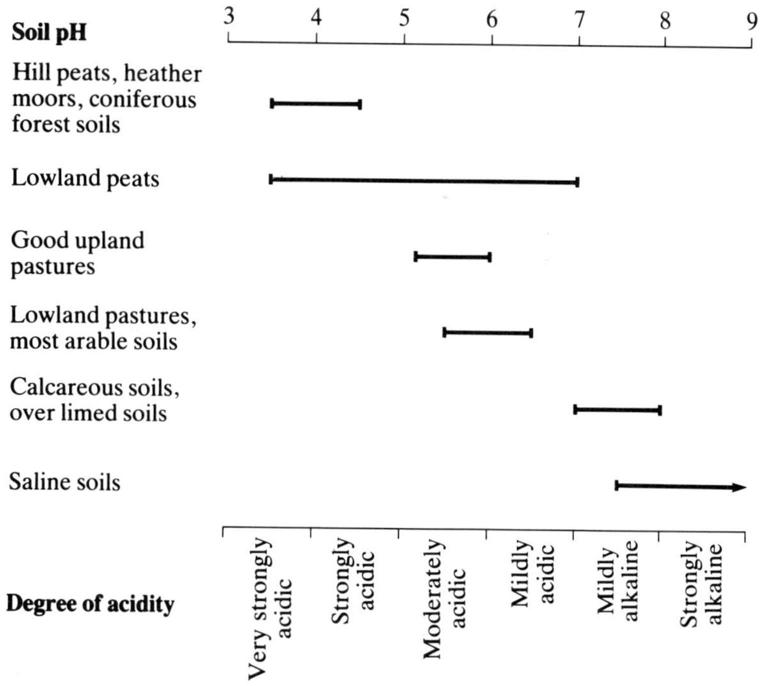

that crops thrive – some, such as oats and potatoes, at the lower end of the range (pH 5.5–6.0) and some, such as barley and sugar beet, at the higher end (pH 6.0–7.0).

The pH values of calcareous soils are 7.0 or greater (up to 8.5). They are controlled by the nature of the chalk, limestone or other parent materials rich in calcium carbonate and are really too high for most crops. There is nothing that can be done economically to reduce the pH of these soils and farmers in vast areas of south-eastern England must use them as they are, taking account of the inevitable fixation of phosphate and the risks of trace element deficiencies. Liming these soils would only exacerbate the problems and must be avoided.

In Britain, however, in common with other parts of the world, many types of soil are derived from non-calcareous parent materials. As a result they are acidic (pH less than 7) and under natural conditions the degree of acidity depends upon the type of parent material, the amount of leaching and the development, if any, of peat on top of the original mineral soil. In Britain the most acidic soils are the strongly leached soils (podzols) and deep peats of the north-west. They represent the extreme acid end of the soil pH range (pH less than 5.0) and cannot support agricultural crops without liming.

A sound liming policy is the key to a good crop growth on all fundamentally acidic soils. Without this the efficiency of manures and fertilizers can be seriously reduced. This was recognized early in British agriculture and, when the Second World War was imminent, the Land Fertility Scheme was launched in 1938. The pH and lime requirement of every field in selected areas were determined. Much of the land was found to be too acidic for efficient crop production and, for the following thirty years, lime was heavily subsidized by successive governments, resulting in outstanding improvements in crop production.

The term lime includes the carbonates, oxides and hydroxides of calcium and magnesium but the carbonates ($CaCO_3$, $MgCO_3$) are much the most widely used, in the form of ground limestones or chalk. These materials are used in preference to other alkalis for neutralizing soil acidity because they are cheap, easily obtainable (there are chalk or limestone deposits in most areas), and easily applied after grinding to a powder. Other alkalis such as sodium, potassium or ammonium hydroxides are both caustic and toxic whereas calcium hydroxide and carbonate are not. Lime applied as ground limestones or chalk persists in the soil for some years, during which time it is gradually converted to calcium and/or magnesium bicarbonate which neutralize soil acids, supply some essential calcium and magnesium to plants and are eventually exhausted through leaching and uptake by crops. Thus judicious repeated applications are needed to prevent renewed acidification of the soil.

Soil pH and lime requirements

The pH values of soils and the quantity of lime required to raise the pH to desirable levels can be assessed fairly accurately by laboratory methods. The national advisory services, as well as some fertilizer and lime merchants, will analyse samples of soil for pH and lime requirement. There are also 'do-it-yourself' pH kits on the market which are based either on colour changes in an indicator solution added to the soil or on readings given by a more robust model of the laboratory pH meter. Provided that the makers' instructions are followed absolutely both methods can give a reasonable guide to the pH of a particular sample. It is critically important, however, to get a representative sample of the soil from a field. This is very difficult and is a job for the trained specialist. The Scottish Agricultural Colleges run such a sampling service but, regrettably, in England and Wales sampling by farmers is encouraged. It is no disrespect to farmers to suggest that they are not expert soil samplers. Nor is it unreasonable to suggest that, in practice, the actual sampling is handed on by them to someone even less expert. This can result in grossly inaccurate results and there is no doubt that sampling and analysis by specialists give very much more reliable results than 'do-it-yourself' methods.

Each soil has a specific lime requirement to bring it to a given pH value. This depends upon its present pH and its cation-exchange capacity which is controlled by the amounts and types of clay and organic matter contained. The greater the cation-exchange capacity the greater the amount of lime required to raise the pH from one value to another. Figure 7.2 shows lime requirement graphs for a low-humus sandy soil (A) and a clay soil containing 10 per cent of well humified organic matter (B). The shaded area (X) represents the range of lime requirement values for the great majority of agricultural soils in Britain. Such graphs may be used to estimate the amount of lime required to raise the pH of any soil from its present level to that desired for a particular cropping

Figure 7.2   Lime requirement, soil pH and soil type.

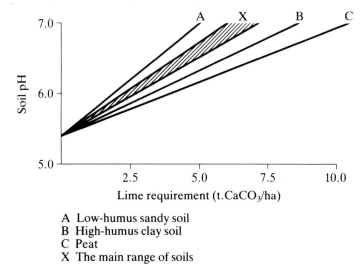

A Low-humus sandy soil
B High-humus clay soil
C Peat
X The main range of soils

system. In Fig. 7.2, 3.5 and 5.5 tonnes $CaCO_3$ per hectare would be needed to raise the pH of soils A and B respectively from 5.3 to 6.5.

Lime requirement graphs are most useful for soils of low pH which are being reclaimed or gradually improved to a more productive state.

Tolerance of acidity and alkalinity by crops

Crops vary considerably in their pH requirements and hence in the amount of lime required to maintain the soil in the 'optimum' pH range. Figure 7.3 shows the optimum range for crops grown on mineral soils in the British Isles and the more extensive range over which the crops will *usually* grow satisfactorily. It is *not* recommended that the extreme ends of the tolerance range should be used. In the optimum range the crop will suffer no acidity problems but there may be a risk of trace element deficiencies at the high pH end of the optimum range on some soils.

Crops grown on peat will usually tolerate more acid conditions than those on mineral soils. For example I have seen excellent crops of healthy scab-free potatoes grown on peat at pH 4.0. This could not happen on a mineral soil. The reason is the relatively

Figure 7.3 Optimum soil pH range for the main crop species.

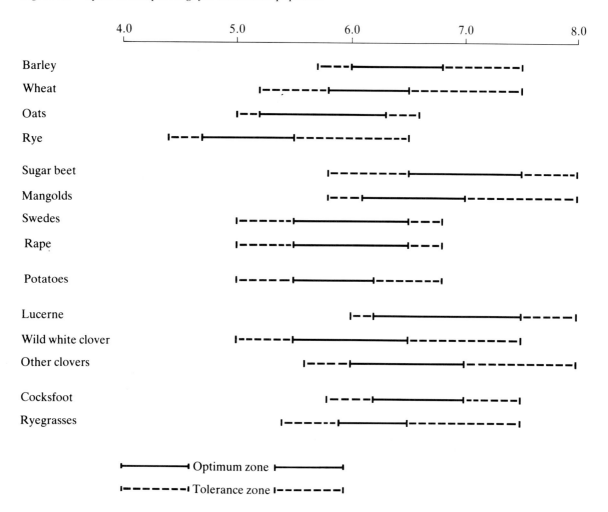

small amounts of potentially toxic elements, such as manganese, present in peat.

Figure 7.3 illustrates clearly some of the problems in deciding the pH range and hence the amount of lime needed for a particular grouping of crops. The problem is not too difficult for poor-land crops, the optimum ranges for oats, rape, swedes and grasses being compatible. In recent decades the problem has also been reduced by the moves towards monoculture, continuous cereal growing or near-continuous cereals broken only by occasional lime-loving crops such as oil-seed rape.

Problems are greatest if, for example, barley and potatoes are to be grown in the same rotation. If sugar beet is also included it becomes impossible to select a pH range that is optimal for all three. The practical solution in such cases is usually to select an optimum pH range for the lime-loving sugar beet and barley (6.5–7.0) and to accept the risks of common scab or trace element deficiencies on potatoes.

Symptoms shown by crops on acid soils

If liming is neglected a time will be reached when crop yields decline and crops show visual symptoms which may be used to diagnose soil acidity. This should *always* be confirmed by soil analysis before liming is undertaken.

Of the cereal crops, oats and rye are acid-tolerant and seldom show symptoms. Wheat and barley plants become dwarfed and develop thin 'spiky' leaves with yellow tips turning brown. The plants may also show reliable symptoms on the roots which become thick and stubby compared with normal roots. They end abruptly, branching little, and have extensive brown scorched areas on their surfaces.

Root crops and brassicas become stunted and many plants die out, leaving areas of bare soil. Leaves tend to 'cup', the margins turning upwards. Yellowing and browning of leaf margins sometimes occurs.

Clovers and other legumes may fail to produce root nodules and plants will wilt rapidly in dry periods. There are no specific symptoms of acidity in grasses but long-term grass/clover leys will lose their clover and the sown species of grass will be gradually replaced by acid-tolerant indigenous species, such as red fescue (*Festuca rubra*), bent (*Agrostis* spp.) and Yorkshire fog (*Holcus lanatus*).

Causes of injury to crops on acid soils

The causes of injury or death of plants in acid soils are not simple. The following points are worthy of note:

- Increased solubility of soil manganese and aluminium as acidity develops can cause toxicity to crops. This is the most common cause of crop failure.
- Little mineralization of organic nitrogen occurs in strongly acid soils and yellowing of plant leaves may result.
- Phosphorus fixation is vigorous at low soil pH values and may result in stunted plants.

● Failure of the injured root system to penetrate large volumes of soil is a major factor and restricts uptake of essential nutrients.
● Certain disease organisms, such as *Plasmodiophora brassicae* which causes 'finger-and-toe' in swedes and clubroot in other brassica crops, thrive under acid conditions and, once established, are not easily eliminated simply by liming.
● Some acidic soils are deficient in molybdenum. This is most likely to affect brassica crops and legumes.
● In soils leached intensively for many thousands of years (sandy soils in wet areas) the availability of copper, zinc and boron may be low because of the long-term reductions in the *total* content of these elements.

Any of these factors can contribute to failure or unthriftiness of crops on acid soils. Injury is usually the result of a combination of two or more of them.

Penalties of underliming

The main penalty of underliming is a reduction in yield, seen first in sugar beet, mangolds, barley and lucerne which thrive only at high soil pH values (Fig. 7.3). For some crops, such as sugar beet, the margin between pH values for complete failure and a successful crop is very narrow, I have investigated a case in which a good crop was growing at pH 5.4 and, in the same field, there were no plants at all at pH 5.2. This should *not* be taken as an indication that sugar beet may be grown at pH 5.4! For other crops such as cereals the margin is wider and there is an insidious decline in yield as the pH gradually falls.

Because of soil variation it is seldom that the whole crop in a field will fail through acidity. There is usually an irregular pattern of failure, partial failure and healthy growth. The first signs of acidity in a field are usually seen on the crests or upper slopes of ridges or hillocks, where the texture is a little lighter because of erosion, and leaching is more rapid than in other parts. Regular patterns of crop failure in strips across a field have usually been caused by poor distribution of lime some years previously. Some alarming patterns can be seen immediately following lime application before cultivations obscure the lines of concentration behind the distributor.

Symptoms shown by crops on mildly acid or alkaline soils

The main symptoms shown by crops grown at pH values of 6.5 or greater are those of trace element deficiencies, notably manganese or boron and less commonly copper.

Most susceptible to manganese deficiency is the acid-tolerant oats crop but barley may also be badly affected. Manganese deficiency may occur on a wide range of crops. On one occasion many years ago when trace element deficiencies were very exciting I was one of a party of fifty or more ecstatic soil scientists taken by staff from the Long Ashton Research Station to visit a garden near Bristol. The hyper-tolerant householder stood by in puzzlement while we observed more than thirty species of vegetables, herbaceous plants and even trees, along with specially planted rows of oats and barley, all showing classical manganese deficiency symptoms. The cause had been gross over-enthusiasm on the part of a previous owner of the house who had heard of the benefits of lime.

The symptom of manganese deficiency on most broad-leaved species is an intensive mottling with very small yellow spots between the vascular network. Even relatively small veins remain green.

In barley numerous dark brown or black oval lesions 1–2 mm long are found along the veins. In oats, larger paler lesions 2–4 mm in length occur, usually pale brown or grey. They tend to concentrate about one-third of the way up the leaf and cause it to collapse.

In contrast to plants short of manganese those deficient in boron seldom show leaf symptoms. Instead there is a breakdown of the cell structure of storage tissue such as the roots of sugar beet and swedes and the heart of celery. The growing points are commonly affected and die out. Cereals and grasses are rarely affected.

Cereals are the main agricultural crops affected by copper deficiency. If the deficiency is severe the leaves tend to die back from the tip and to take on a spiral form. This condition is described as 'wither-tip'. Another very serious and frustrating symptom occurs when the cereals make apparently normal vegetative growth and form ears but no grain develops. The term 'blind-ear' is applied to this form of the disease.

Penalties of overliming

The main penalties associated with excessive applications of lime arise from induced trace element deficiencies. Unless treated promptly after the appearance of symptoms, manganese deficiency may reduce crop yields severely. Remedial treatment is essential. The most successful method is to spray with solutions of manganese sulphate (Table 15.8, p.223).

Crop losses caused by boron deficiency are more insidious. By the time the first symptoms have been observed it is usually too late to apply remedial treatments. Prevention in future crops by such measures as the use of boronated fertilizers (Table 15.8) is normally the only recourse.

Copper deficiency is less common but, if left untreated, can reduce the yield of oats or barley to as little as 0.5 t/ha. Some improvement in crops showing early 'wither-tip' symptoms can be achieved by spraying with solutions of copper oxychloride but symptoms of 'blind-ear' appear too late for spray treatment to be of any value (Table 15.8). Long-term treatment should be undertaken for subsequent crops by application of copper sulphate ($CuSO_4.5H_2O$) to bare soil.

Effects of overliming commonly last for 10–15 years. Such applications are unnecessary, inefficient and costly.

Liming materials

Many of the liming materials formerly used in large quantities are now only rarely used. They include burnt lime (CaO), slaked or hydrated lime ($Ca(OH)_2$) and waste limes from sugar beet and other factories. They have no advantages over the ground chalks or limestones which have replaced them and they are unpleasant to use. It is interesting and perhaps perplexing to look back on the era between the two World Wars when burnt and hydrated limes dominated the UK agricultural market. Limestone was burnt and then sometimes slaked and dried to a powder by its own heat of hydration and sold in bags. More commonly the burnt lime in large lumps was carted directly on to the field and allowed to slake and steam in large heaps and spread from the heaps, commonly by hand shovel. The result was appallingly bad distribution, overlimed and underlimed patches and, above all, gross excesses of lime where the heaps had been, giving rise to roughly circular areas of manganese deficiency in oats or boron deficiency in sugar beet. In retrospect it is puzzling to think of the vast expenditure of energy used to burn the limestone followed by applying it to the soil where it is gently converted back to calcium carbonate. It was not, however, until the late 1930s that the now widely used ground limestones began to replace the burnt and hydrated limes.

The specifications of liming materials are legally controlled. The seller is required to quote the quality of the lime in terms of calcium oxide or calcium carbonate equivalent. The fineness of grinding must also be stated within narrow limits. All the material must pass through a sieve with approximately 3 mm holes and at least 40 per cent must pass through a standard 100 mesh sieve (approximately 0.15 mm apertures), thus ensuring a high proportion of fine, quick-acting material.

The farmer's choice of lime is often limited by the geographical distribution of chalk and limestone deposits. Much liming is done by contractors and they will usually buy from the cheapest source of good material. Most areas of the British Isles are close to a source of lime that is currently being worked and transport costs dictate that local chalk or limestone will be offered to the farmer. Most materials on offer, irrespective of type, are similar in neutralizing value, equivalent to 85–95 per cent $CaCO_3$. In areas where there is a choice of sources there is a slight advantage in speed of action of chalk or the softer types of limestone such as oolite over the harder carboniferous limestones but this does not justify paying a premium for them.

The special case of magnesian limestone

Most limestones as well as chalk consist mainly of calcium carbonate. The exception is magnesian limestone, sometimes described as dolomitic. The Fertilizer and Feeding Stuffs Regulations accept as 'magnesian' any limestone containing more than 3 per cent Mg (11 per cent magnesium carbonate) but magnesian limestone in its purest form is dolomite ($CaCO_3.MgCO_3$), which contains by weight 54.3 per cent $CaCO_3$ and 45.7 per cent $MgCO_3$. There is a narrow belt of magnesian limestone which approaches this composition stretching from north-east England to the Midlands and a similar source in north-west Scotland.

Many areas of the British Isles have soils that are by no means rich in magnesium. It is a major plant nutrient neglected in compound fertilizers. The magnesium content of crops used for stock feed is depressed both by the use of high rates of potassium fertilizer and also by the routine use of commonly available chalks or calciferous limestones with a high ratio of calcium/magnesium. The serious and widespread occurrence of hypomagnesaemia (low blood magnesium) and the deaths of ruminant livestock from hypomagnesaemic tetany in the 1950s pointed up this problem.

The regular use of magnesian limestone, in preference to calciferous limestones, either for every application or for alternate applications, will help to alleviate magnesium deficiency in both crops and stock. For this purpose it is essential to use a magnesian limestone containing much more magnesium than the 11 per cent $MgCO_3$ defined as the minimum legal level. A good sample would contain some 50 per cent $CaCO_3$ and 40 per cent $MgCO_3$. This material is equivalent in neutralizing value to most calciferous limestones. It is not available locally in many areas but it is well worth while paying a 20 per cent premium for its transport. It is a very cheap source of magnesium compared with any other magnesium fertilizer.

It is important, however, to stress that although liming with magnesian limestone will usually prevent magnesium deficiency in crops it cannot, *alone*, be relied upon to prevent problems in stock. Dosing of stock directly with magnesium compounds is commonly the only effective remedy for hypomagnesaemia.

Liming policy

Liming policy should aim to:
● Raise the soil pH into the range which is optimal for the crops to be grown (Fig. 7.3).
● Maintain the pH in that range by regular small applications of lime (4–6 t/ha of $CaCO_3$ or its equivalent) at intervals of 1–5 years depending upon the soil type, rainfall, fertilizer use and intensity of cropping.
● Avoid excessive lime applications, thereby restricting leaching losses of lime and reducing the risk of trace element deficiencies.

Inadequate liming can have catastrophic effects on crop yields. None the less, in times of financial stress, liming is one of the first essential operations to be neglected. This is easy to understand because the effects of neglect are not obvious for a year or two and the temptation will be to buy immediately rewarding foodstuffs or fertilizers instead. Strong evidence of this comes from the liming statistics for 1970–80. As the decade progressed less lime was applied and by 1980 insufficient was being used on a UK basis to *maintain* the pH of agricultural soils. This resulted partly from the removal of government subsidies which had done so much to encourage liming and partly from the economic state of farming. It cannot be stressed too strongly that a good liming policy is essential.

Raising the soil pH

On newly reclaimed acidic soils or soils on which liming has been neglected for more than five years the first problem is to raise the pH to the required value. This would also apply if it is intended to introduce lime-loving crops to a rotation previously consisting of acid-tolerant crops, especially if at the same time there is more intensive cash cropping at higher yields.

In all such cases analysis of a well-taken soil sample for pH and lime requirement by a reputable organization is essential. Using graphs such as those in Fig. 7.2, a reasonable estimate of lime requirement can be made. It will vary from nil in calcareous soils to very high values in excess of 10 t/ha of $CaCO_3$ in some strongly acidic soils (pH 3.0–4.0) to be reclaimed.

With such high lime requirements, fortunately now rare, no attempt should be made to raise the pH to the level eventually required with a single application. Applications of ground limestones at rates greater than 7 t/ha are both wasteful, because excessive amounts of lime will be leached, and dangerous because of uneven distribution leading to irregular areas of trace element deficiency.

Maintenance liming

Once the pH value of the soil has been raised to the optimum level for the crops to be grown it is necessary, because of losses of calcium and magnesium by leaching and offtake by crops, to use regular maintenance dressings of lime. Traditionally this has been done by the use of a standard application, commonly 5.0–5.5 t/ha of ground limestone, at intervals of 3–4 years on rapidly leached sandy soils and 5–6 years on high-humus clay soils.

Figure 7.4   *Soil pH and time after liming.*

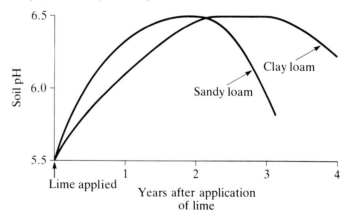

Figure 7.4 gives an example of the changes in soil pH over time after an application of lime. Lime acts slowly in the soil and the time taken to reach the maximum pH value is usually about 1.5–2 years. The period over which this maximum pH is maintained depends upon the rate of leaching and to a much lesser extent

upon the amounts of calcium and/or magnesium being removed in crops (equivalent to 50–200 kg $CaCO_3$/ha per year). Under former cropping and fertilizer systems the period was some 2–4 years, after which the pH began to fall. Maintenance liming was needed either immediately before or very shortly after this, hence the tradition of liming at 4–6 year intervals. Until recent years the amounts of calcium and magnesium leached depended very largely on the excess of rainfall over 'transpiration plus evaporation', particularly in the winter months. During the last thirty years, however, the amount of nitrogenous fertilizers used in UK has increased greatly. The soil acidifying effect of these fertilizers, along with the increased removal of calcium and magnesium as crop yields increase, has necessitated *either* larger applications of lime at the traditional 4–6 year intervals *or* more frequent liming at the current rate of approximately 5 t/ha $CaCO_3$.

The acidifying effect of some nitrogenous fertilizers Nitrogen fertilizers which contain ammonium compounds acidify the soil. Other nitrogen compounds such as urea and ammonia are rapidly converted to the ammonium form in the soil and therefore also cause some acidification. The acidity is brought about by the release of hydrogen ions into the soil during the bacterial conversion of ammonium to nitrate. Theoretically, every kilogram of ammonium nitrogen applied to the soil can cause the loss of 3.6 kg of $CaCO_3$ from the soil. The conditions required for such an extreme loss exist only in bare soil. Growing plants reduce the acidifying effect by taking up some of the ammonium directly and thereby preventing the nitrification process. The degree to which plants reduce the acidification brought about by ammonium fertilizers will obviously vary a good deal from crop to crop. Top dressings to rapidly growing grass will be readily taken up by the crop before nitrification and thus acidification will be reduced. In contrast, ammonium fertilizers applied before planting crops such as potatoes, which grow slowly in the early season, will have a much greater acidifying effect.

It is, therefore, impossible to predict accurately the amount of calcium carbonate lost in a single season as a result of the use of ammonium fertilizers. A good rule-of-thumb is to assume that, over several years, 50 per cent of the ammonium applied will be taken up before nitrification. The other 50 per cent will cause acidification.

On this assumption 1 kg of NH_4 nitrogen applied will bring about a loss of 1.8 kg $CaCO_3$.

Nitrate fertilizers do not cause soil acidification. In fact some, such as calcium nitrate, tend to make soil more alkaline.

Table 7.1   Acidifying effects of some nitrogenous fertilizers.

Fertilizer	Comparative acidifying effect
Ammonium nitrate/calcium carbonate:	
15% N	Nil
23% N	*
26% N	**
Ammonium nitrate	***
Urea	***
Ammonia	***
Ammonium sulphate	*****

The fertilizers are compared in terms of calcium carbonate loss per kg of fertilizer nitrogen applied in standard conditions. The greater the number of asterisks the greater the acidification.

Table 7.1 shows the relative acidifying effects of different nitrogenous fertilizers. The relatively small acidifying properties of the ammonium nitrate/calcium carbonate fertilizers is a result of the in-built neutralizing effect of the calcium carbonate component. As the nitrogen content of this type of fertilizer is increased the amount of calcium carbonate contained becomes insufficient to neutralize the acidity produced by the ammonium nitrate.

The acidifying effects of nitrogenous fertilizers have become much more important in recent years as new crop varieties, especially cereals, have permitted the use of much greater amounts of fertilizer nitrogen. The approximate total amount of nitrogen applied per year in fertilizers in the UK increased five-fold from 250 000 tonnes in 1950 to 1 250 000 tonnes in 1980 – i.e. by 1 million tonnes. This would be roughly equivalent to 1 800 000 tonnes of ground limestone – twice the annual amount of lime applied in the whole of Scotland.

These effects are so important that they are now included in the Agriculture Regulations for calculations of lime and fertilizer residues, used when a farm changes hands to assess the value of recent applications for the purpose of compensation.

Frequency of liming   Figure 7.5 relates the annual loss of calcium carbonate from soils with excess winter rainfall and average annual application rates of fertilizer nitrogen. It takes account of losses in drainage as well as calcium and magnesium taken up by crops. There will obviously be some variations from the graph presented. For example, lime losses from a very sandy soil would be greater than those shown.

Figure 7.5   *Effects of excess winter rainfall and applications of nitrogenous fertilizers on the annual loss of calcium carbonate from soils. Devised from data from many sources.*

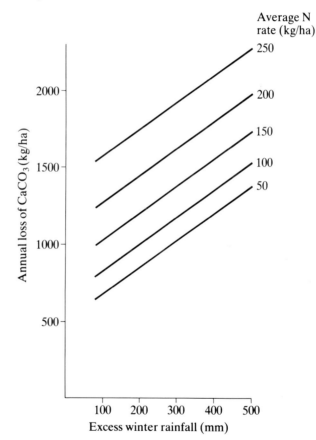

None the less Fig. 7.5 may be taken as a good guide to annual lime losses in British agriculture.

The important thing is to recognize that if nitrogen is being used at moderate (100 kg/ha per year) or high rates (300 kg/ha per year), averaged over some years, acidification will be quite rapid. Take for example a soil with a modest excess winter rainfall of 350 mm, at an average input of 100 kg N/ha per year: this soil would lose some 500 kg $CaCO_3$ each year. With an excess winter rainfall of 500 mm and a nitrogen input of 180 kg/ha, annual losses of $CaCO_3$ would reach 1 tonne. Such rates of loss suggest that a new approach to the *frequency* of liming is required. In some cases, especially where long-term grass is grown or minimal cultivation arable systems are used and the soil is not turned over, leaching is so rapid that the important surface 5–10 cm can become much more acid than underlying soil and surface roots may suffer from acid damage.

The old adage 'little and often' has always applied in liming practice and is even more apt now that leaching has been enhanced by modern farming systems. The alternative of larger applications at the traditional 4–6 year intervals is unattractive because it leads to more leaching and the risk of trace element deficiencies.

Table 7.2   Maintenance liming–interval in years* required between lime applications† as affected by excess winter rainfall and average annual rates of nitrogenous fertilizers.

Excess winter rainfall (mm)	Average annual rate of N fertilizer (kg N/ha)				
	50	100	150	200	250
100	7	6	5	4	3
200	6	5	4	3	3
300	5	4	4	3	2
400	4	4	3	3	2
500	4	3	3	2	2

* The intervals in this table are calculated from the data in Fig. 7.5.

† It is assumed that a standard rate of 5 tonnes of high-quality ground limestone will be used.

Table 7.2 gives examples of suitable maintenance liming programmes based on the same criteria as Fig. 7.5. It is assumed that the soil is already in the desired pH range. Note that the *frequency* of liming rather than the *amount* applied at a dressing has been increased as leaching increases.

Table 7.2 should be regarded as a simple guide to maintenance liming and it would be very wise to have the soil sampled and the soil pH checked by a reputable laboratory at 4–5 year intervals or even more frequently if your farming system is intensive.

Timing of lime applications is not so important now as it was when lime levels were being gradually raised. One general rule should be followed. The tendency is to apply lime during the winter or early spring before the most demanding crop of the rotation is grown. This is too late for the lime to be fully effective for that crop. Lime should be applied fully one year earlier, that is not in the previous winter but the one before that. This rule becomes unimportant as the frequency of liming increases, in which case the timing of applications becomes a matter of convenience. It should be remembered that the passage of heavy liming vehicles can seriously damage the structure of wet soils and cause pan formation.

Surface application of lime followed by shallow cultivation is satisfactory provided that the distributors give a reasonably uniform spread across the field. It is seldom necessary except when reclaiming very acid soils to use split dressings. The first ploughing after liming will ensure a reasonable vertical distribution of the lime.

Liming practice in the future

Table 7.2 illustrates that we are, in some instances, only a step away from *annual* applications of lime. This would be quite a reasonable proposition at rates of 0.5–1.2 t/ha of calcium carbonate, based on annual $CaCO_3$-loss calculations. It would be most applicable to intensive crop production systems. This proposal could be helped along by the delivery of lime to the farm in 'big bags' similar to those now used for fertilizers. It is even conceivable that the finely ground limestone could be granulated in the interests of ease and uniformity of spreading but obviously the farmer would have to pay a premium for this.

We are at present committed to the use of ground limestone, 40 per cent of which passes through the very fine 100 mesh sieve. The use of this very fine material was amply justified when the urgent need was to raise soil pH values and the legislation controlling fineness of grinding was made to fulfil this need. The current need, however, is to *maintain* pH values. It is very likely, but not yet proven, that this could be done by using much more coarsely ground limestone – for example, the bulk of the material being 1–3 mm in size with some 20 per cent passing a 1 mm sieve.

To adopt such a material would need a change in the present Fertilizer Regulations. The savings of energy used in grinding would be great and the cost of liming should, as a result, be reduced.

Chapter 8   Manures

Manures are waste plant and animal products which are recycled by returning them to the soil. This may occur directly in the excreta of grazing animals but in the case of housed or confined animals the excreta need to be processed or stored before they are spread on the soil. Crop residues such as stubble, straw, plant roots and potato haulms, returned directly to the soil also have a manurial value. Some crops are grown especially to plough in directly as 'green manures'.

Farmyard manures and slurries are regrettably regarded by many farmers simply as unavoidable by-products of their farming system involving major storage and disposal problems. As a result they are seldom well conserved and are commonly used on convenient areas for crops which are not responsive.

Recent advances in dung-spreading machinery have made distribution in the field more uniform and infinitely easier than the old cart-and-fork systems. There are, however, still large variations from point to point in a field of the nutrients received by the soil from manures as compared with fertilizers.

Despite these problems, well-made, carefully conserved, evenly spread and immediately incorporated farmyard manures or slurries can improve soil fertility both in the first year after application and in the long term.

The functions of manures

Manures have two main functions – to supply nutrients and to supply organic matter.

Manures as a source of nutrients

The concentration of nutrients in manures is very low compared with that in fertilizers. One tonne of traditional farmyard manure made with straw supplies N, P and K in similar amounts to only 50–100 kg of a modern concentrated compound fertilizer. However, manures also supply useful amounts of calcium, magnesium, sulphur and trace elements (see Table 8.2, p. 90), all largely neglected in modern fertilizers. If well conserved and incorporated into the soil they can give considerable savings in the amounts of fertilizer required.

Appreciable proportions of the total nutrient content of manures occur in complex organic forms which have to be mineralized before they release available nutrients. Thus, not all will be available for the first crop after application. In Table 8.1 estimates are given of the quantities of nutrients that will be available from each main type of manure for the first crop after application. The total amounts of N, P_2O_5 and K_2O are also given. The nutrients remaining after the first crop may become available for later crops although it is not possible to predict precisely when or in what quantities this will occur.

Manures as a source of organic matter

Manures are, by nature, organic. Their organic matter is attacked and transformed by micro-organisms when returned to the soil. Much of the carbon is converted to carbon dioxide and makes no long-term contribution to the organic matter content of the soil. Other parts of the organic matter are converted to humus, a black or dark brown, colloidal, very complex organic material which remains in the soil. Humus is a very valuable soil component which increases the ability to hold water available to the plant and, through its very high cation-exchange capacity, reduces the leaching of nutrients.

All manures make some contribution to long-term soil fertility and the maintenance of humus in the soil but the *extent* of this contribution should not be overestimated. In fact, very large amounts of manures need to be applied to have significant long-term effects on the organic matter content of the soil. The main reasons for this are the very high water content of many manures and the loss of much of the organic matter during decomposition in the soil. Even bulky straw-based farmyard manures contain about 75 per cent of water and slurries more than 90 per cent so that 1 tonne of these manures will add only 250 kg or 100 kg of organic matter to the soil respectively. Figure 8.1 shows the fate of 40 tonnes of farmyard manure applied to a fertile soil.

It is reduced to some 2.5 t/ha after humification is complete. This process takes only a few months or, at the most, a year or two. The final residues represent only 0.1 per cent of the weight of 1 ha of soil to a depth of 25 cm. Slurries, which contain a larger proportion of material easily metabolized by bacteria, will make

Figure 8.1   *The fate of farmyard manure in soil.*

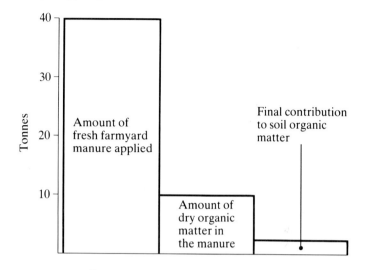

an even smaller contribution to the organic matter content of the soil. Thus any attempt to increase the soil organic matter simply by applying farmyard manure requires regular *annual* dressings of 40 t/ha or more. This is simply not feasible if only the farmyard manure produced on the farm is used. Typically, in order simply to maintain the humus content of a sandy loam soil, annual applications of 15–25 t/ha of farmyard manure would be needed.

Thus, in normal farming practice the use of bulky straw-based farmyard manures can be expected to make only a small but useful contribution to the quantity of humus in the soil. Slurries will contribute less.

Types of manure

Manures may be classified as:

- *Traditional farmyard manures* – bulky solid products in which straw or other organic materials have been used to absorb liquid excreta from animals.
- *Liquid manures* – usually known as slurries, consisting of excreta which have been deposited on solid or slatted floors, with no straw, and then washed into lagoons or storage tanks.
- *Processed organic materials* – produced off the farm and applied to the land. This group includes sewage sludge, town refuse and seaweed.

● *Crop residues returned directly to the soil* – straw, stubble and roots of cereals and grasses, turnip leaves and other unwanted parts of plants, and green manure crops grown specially for ploughing in.

Traditional farmyard manures

Most of these bulky manures are based on straw or other materials with a high carbon–nitrogen ratio such as wood chips, sawdust or peat, all used as bedding for animals. For these materials to be effectively converted to organic manures they must be rotted – partly decomposed by fungi and bacteria. This happens rather slowly if they are incorporated directly into the soil but much more quickly in farmyard manure. The process requires much more nitrogen than that contained by the straw or other bedding materials to enable the micro-organisms involved to live and multiply. This nitrogen is supplied by the urine and faeces of the animals and this triggers off the decomposition which starts during use as bedding, continues in storage and ends in the soil, leaving a residue of humus.

Fermentation

The microbial decomposition of straw, dung and urine starts as an aerobic process in the loose litter where oxygen is plentiful. During this rapid phase of decomposition a good deal of heat is generated, contributing to the pleasant atmosphere of enclosed cattle courts. Much carbon is lost as carbon dioxide and the urine nitrogen, mainly urea, is converted to ammonia. Some of this is lost into the atmosphere, some absorbed by the decomposing mass and some converted to the easily leached nitrate.

As the animals compact the bedding and the micro-organisms exhaust the supply of readily available oxygen, anaerobic decomposition takes over. During this phase the material decomposes much more slowly and the temperature falls. Complex nitrogen compounds are produced and preserved until the manure is applied to the soil.

Some losses of valuable nutrients, particularly nitrogen, are inevitable during the fermentation or 'making' of farmyard manure. They may be reduced in several ways:
● Reduce the period of aerobic fermentation by ensuring adequate compaction through trampling by stock.
● Reduce or prevent leaching by using solid waterproof floors in the living quarters of the stock.

● Add fertilizer phosphate to the manure at weekly intervals during the housing period. This will not only help to reduce ammonia losses but will increase the phosphorus content of the final product. It is preferable to use a water-soluble phosphate such as triple superphosphate and the phosphorus from these sources will be readily available for the first crop after spreading the manure. The amounts used should be equivalent to $0 \cdot 15 - 0 \cdot 25$ kg P_2O_5/day per animal in the court. Water-insoluble ground mineral phosphates may be used but at double the P_2O_5 rate. Ammonium phosphates should not be used as they will not absorb ammonia from the fermenting mass.

The addition of phosphates to farmyard manure, although widely practised in parts of Europe and the USSR, is almost completely neglected in the British Isles. It is easily done, reduces nitrogen losses appreciably and increases the normally low available phosphorus content of the manures to more satisfactory levels.

Storage
If feasible, as in the gradually-filling Scottish cattle courts, it is most satisfactory to store the manure where it is made until it can be spread. Storage systems using roofed, rectangular waterproof pits, as in some European countries, are also satisfactory.

If this is not possible, the 'made' manure may be stored in deep piles (2 m high) in the open air, preferably on a waterproofed concave cement base or on previously puddled clay so that the liquids are not lost. A simple polythene cover stretching beyond the solid base will reduce leaching losses resulting from rain.

Such precautions are well worth while. As much as one-fifth of the nitrogen and one-third of the potassium content can be lost as a result of inadequate storage. Such liquids as are lost become serious pollutants of streams and other waterways because of their high BOD (biological oxygen demand) which can deprive stream life of oxygen. It is now illegal to allow such pollution to occur.

Direct application from cattle court to soil is obviously efficient in terms of labour and energy and will ensure minimal storage losses of nitrogen and potassium. Direct application is not, however, always feasible and, unless manure so applied can be rapidly cultivated in, there will be nitrogen losses by volatilization of ammonia.

Application and use
Application should give as even a distribution as possible over the land surface in as finely divided a form as can be achieved. Most

modern dung-spreading machines do this quite well, although all too often one sees large dollops of dung flying through the air and landing with a disconcerting clop. The machines are generally cylindrical, breaking up and throwing out the dung by means of rotating blades.

The fineness of division, so desirable for even distribution, unfortunately makes for further losses of nitrogen, especially in hot windy weather. The aim should therefore be to spread the well-rotted dung on bare soil or stubble and to incorporate it immediately.

Traditional rates of application are 25–40 t/ha. They have evolved from practical experience and certainly the benefits of smaller, more frequent applications are doubtful because of difficulties of even distribution and the inefficient use of machinery.

Allowing 10 tonnes of fresh manure per head of cattle per housing period in a loose housing system, using a ratio of 1 : 20 *by weight* of straw : faeces + urine and assuming losses of 35–45 per cent during making, 100 cattle would produce 600–700 t/year of manure. This would be sufficient for some 24–28 ha each year, one fifth of a 120–140 ha farm, at a modest rate of 25 t/ha.

Response of crops

Crops vary considerably in their response to farmyard manure applied during the previous winter or spring. The most profitable yield increases usually occur in crops of potatoes, sugar beet, mangolds, turnips and vegetables. On mixed farms the manure should certainly be reserved for these responsive crops, and especially for potatoes. Many series of field experiments have been done on yield responses to farmyard manure applied at rates of 25–30 t/ha. Increases in sugar beet yield vary from 0.8–4 tonnes of fresh beet/ha. Potatoes are even more responsive, yield increases of 7–13 t/ha being found from farmyard manure alone.

Unfortunately the present trends in agriculture are to concentrate the crops which respond best to farmyard manure in intensive arable areas where less and less stock are kept. There is also a trend towards continuous or intensive cereal production and even stock farms with a small proportion of arable land are tending more and more towards cereal production. These trends lead inevitably to the less effective use of farmyard manure for cereals or grassland. Cereals are not nearly so responsive as potatoes and root crops but farmyard manure is particularly ineffective on established grassland because of the impossibility of incorporation and consequent losses of nitrogen. Application to grass also contaminates the herbage and renders it unpalatable to animals for some days or weeks.

Fertilizer equivalents

Table 8.1   N, P_2O_5 and K_2O equivalents of manures made on the farm.

	Available for first crop after application			Total		
	Nutrients in kg/tonne					
	N	P_2O_5	K_2O	N	P_2O_5	K_2O
Farmyard manure (cattle)						
Range	1.5–2.5	1.4–1.6	2.0–5.0	5.0–7.0	2.5–3.5	4.5–7.0
Average	2.0	1.5	3.0	6.0	3.0	5.0
Farmyard manure (cattle, supplemented with phosphate)						
Range	1.5–2.5	2.5–3.5	2.0–5.0	5.0–7.0	5.5–6.5	4.5–7.0
Average	2.0	3.0	3.0	6.0	6.0	5.0
Poultry manures:						
Dried	25	15	17	40	30	25
Deep litter	10	10	10	17	18	13
Broiler litter	15	12	12	24	20	15
	Nutrients in kg/1 000 litres					
Slurries: Cattle (undiluted, 10% dry matter)						
Range	2.0–3.0	0.8–1.2	3.5–5.5	4.6–6.0	1.7–2.4	5.0–7.5
Average	2.5	1.0	4.5	5.0	2.0	6.0
Pig, according to feeding system						
Dry meal (10% dry matter)	4.5	2.5	2.5	6.5	4.5	3.0
Liquid (6–10% dry matter)	3.2	1.5	1.5	5.0	2.5	2.0
Whey (2–4% dry matter)	1.8	1.0	1.4	2.8	2.0	1.8
Poultry (undiluted, 25% dry matter)						
Range	8.0–10.5	4.0–6.0	4.0–6.0	12.0–15.5	8.5–11.5	6.0–7.5
Average	9.5	5.0	5.0	14.5	10.5	6.5

Source: Data from several sources.

Table 8.1 gives approximate values for the quantity of nutrients available for the first crop after application and also the total amounts of nutrients contained.

The lower end of the range of values given will apply to poorly made and poorly stored manures subject to excessive leaching and volatilization losses and to cases where losses after application are high. The upper part of the range applies to manures made and stored under ideal conditions, applied shortly before incorporation and followed immediately by a responsive crop.

The *average* amounts of nutrients available for the first crop are useful in calculating savings that can be made in fertilizer rates. Note that they are equivalent to only one quarter of the *total* nitrogen content of the manure but that rather greater proportions of phosphorus and particularly potassium are available for the first crop. Some of the remaining N, P and K will become available to succeeding crops but it is not possible to forecast precisely in what amounts. They must be regarded simply as a small bonus for future crops.

A very general guide to the compound fertilizer equivalent is that: 25 tonnes of farmyard manure = 500 kg of 10 : 7.5 : 15 fertilizer. If phosphates have been added during the making of the manure at the rate of 3 kg P_2O_5/t of final product: 25 tonnes of 'phosphated' farmyard manure = 500 kg of a 10 : 15 : 15 compound fertilizer.

Table 8.2   Elements other than N, P and K supplied by farmyard manure.

Major elements	(kg/tonne)	Trace elements	(g/tonne)
Calcium	7–10	Manganese	50–100
Magnesium	2–3	Zinc	20–40
Sulphur	2–3	Boron	10–15
		Copper	10–12
		Cobalt	0.8–1.2
		Molybdenum	0.4–0.7

Source: Data from several sources.

In addition to nitrogen, phosphorus and potassium, farmyard manure contains the full range of other elements needed by the plant (see Table 8.2). The actual amounts present are, as for N, P and K, very variable and not all will be available for the first crop after application. None the less farmyard manure is a very useful source of calcium, magnesium, sulphur and the trace elements, all of which are neglected in most modern fertilizers. Table 8.2 gives the likely range of these elements in well-made farmyard manure. *Annual* applications of 25–35 t/ha of such manure would probably more than balance the losses of these elements from the soil.

Other farmyard manures

Very little traditional farmyard manure is made from species other than cattle. Some is made from pig excreta and is much less objectionable than pig slurry. It may be regarded as similar in action to cattle manure but is twice as rich in phosphorus and only half as rich in potassium.

Thus 25 tonnes of *pig* farmyard manure = 500 kg of a 10 : 15 : 7.5 compound fertilizer.

Poultry manures
Deep litter and 'broiler' house manures

Deep litter for poultry usually consists of chopped straw or wood shavings in a layer 10–15 cm deep. When the excreta are added the litter becomes moist but remains aerobic – similar conditions to the early stages of cattle manure fermentation. Aerobic fermentation occurs with the production of heat and the loss of some carbon dioxide and ammonia.

In broiler houses, the litter is changed more frequently and there is less ammonia lost because of the restricted amount of decomposition. This results in a manure rather richer in nitrogen than that from deep litter houses. Both types are reasonably dry and well-conditioned. They still contain much undecomposed straw or wood shavings which will decompose slowly if the manure is stored or added to the soil. The approximate NPK equivalents in kg/t are given in Table 8.1.

Dried poultry manure

Some poultry units have drying plants which rapidly dry out raw excreta or broiler litters until they contain less than 15 per cent water. This minimizes ammonia loss and produces a rich well-conditioned material, with little offensive smell, that can be handled in fertilizer distributors, although required rates of application will be high and the material is much less dense than ordinary fertilizers. The NPK equivalents of dried poultry manure are given in Table 8.1.

Liquid manures (slurries)

Slurries are made from the excreta of cattle, pigs and poultry. They are, essentially, animal excreta diluted with the water used to clean out living areas. No bedding material is added, the stock being kept on solid or slatted floors. The excreta are swept, scraped or washed into large storage lagoons or tanks and kept until required or until the limited storage capacity makes disposal essential.

Fermentation and storage

The microbial actions during slurry storage are quite different from those involved in making traditional farmyard manures. There are two main reasons for this. Conditions are mainly anaerobic because of the rapid exhaustion of oxygen in the liquid medium. This can be alleviated by stirring and bubbling air through the slurry but most farmers are content to leave the slurry undisturbed for long periods. Because of the high specific heat of water, which makes up 90 per cent or more of slurry, temperature rises during fermentation are only small. The result is slower, less complete decomposition of the organic matter and smaller losses of ammonia in the early stages than in farmyard manures.

Provided that the storage tanks are waterproof, losses of phosphorus and potassium are negligible. Most of the urea from the urine is eventually converted to ammonia, much of which is retained as a result of its high solubility and the volume of water present. Some loss will occur, most rapidly in warm weather, from slurry which is agitated and aerated in tanks with a large surface area of liquid exposed to the atmosphere. If not aerated during storage much of the organic matter remains as evil-smelling products of anaerobic decomposition and this causes smell pollution problems but these are slight in cattle slurries as compared with the stench of unaerated pig slurry.

Methods of making and storing cattle, pig and poultry slurries are essentially the same. Special care must be taken in storing poultry slurry, because of its high nitrogen content, to avoid ammonia losses by enclosing the slurry tanks.

In all slurry management systems, however, the main problem is to restrict the amount of water used in cleaning to a minimum and to avoid the inflow of rainwater to the tanks. This is essential to reduce carting and spreading costs and damage to soils during spreading.

Separation

In recent years apparatus has been developed to separate large proportions of the solid material *before* storage. If the water content of the separated solids can be reduced to less than 82 per cent they will not flow and can be stored and used in the same way as farmyard manure although the product will be different in composition because of the lack of bedding material. It will, for example, have a much lower carbon–nitrogen ratio and will be more rapidly mineralized on contact with the soil. Some separators can take out a solid fraction as dry as an average farmyard manure (25–30 per cent dry matter) and this can be added to the normal manure heap, enriching the product.

The liquid part can be stored in normal slurry tanks. The main advantages of separating the liquid lie in the more uniform density of the material and its free-flowing properties, permitting easy pumping. It flows and spreads easily and is less likely to block spreading machines than unseparated slurry. Most separated liquids contain 5–7 per cent dry matter and are therefore less likely than unseparated slurries to 'cap' the soil with a carpet of solids.

Fertilizer equivalents of slurries

Cattle slurry As with farmyard manure, the composition of cattle slurry is very variable, depending on the type of animal, its feeding regime and the way in which the slurry is managed. An added variable not applicable to farmyard manure is *dilution* with water. The degree of dilution is very important in estimating the fertilizer equivalent of the slurry.

The amount of undiluted slurry produced by cattle per day is roughly proportional to their body weight. A 500-kg dairy cow will produce 35–45 litres of urine plus faeces per day; a 350-kg bullock, 25–35 litres. The amount of dilution from washing water can be estimated from these figures, using the ratio of expected output of undiluted slurry to the actual volume in the tank as measured by dipstick or graduations on the side of the tank.

Table 8.1 gives the range of nutrient equivalents of *undiluted* cattle slurries in terms of 'total' NPK and amounts estimated to be available for the first crop. From Table 8.1, 25 000 litres (25 tonnes approximately) of *average undiluted slurry* will be approximately equivalent to 500 kg of a 12.5 : 5 : 22.5 fertilizer for the first crop. However, nitrogen losses from slurry applied in autumn and early winter and left on the soil surface can reduce the amount of nitrogen available to the crop to as little as one-fifth of that potentially available. It is *essential* to incorporate slurry immediately for maximum effectiveness.

The two fractions of separated slurries are surprisingly similar in nitrogen content but there will be a slightly greater proportion of potassium in the liquid part and slightly more phosphorus in the solid fraction than in the original slurry.

In addition to their NPK value slurries will supply calcium, magnesium, sulphur and trace elements in approximately similar proportions to those in farmyard manure (see Table 8.2). Slurries will contain less of these elements than farmyard manure because of their greater water content.

Pig slurry   The composition of pig slurry is even more dependent upon methods of feeding than cattle slurry but, because of the close control of feed and drink in pig units, is more predictable. Approximate composition according to feeding methods is given in Table 8.1.

Dry meal feeding or liquid feeding with a low water–meal ratio of around 2.5:1 gives slurries with approximately 90 per cent of water. With a water–meal ratio of 4:1 the slurry will contain some 94 per cent of water. Slurry from whey-fed pigs is even more dilute (94–97 per cent water). This is the main factor controlling the analysis of pig slurry. It also has a major influence on the *amount* of slurry produced (4 litres/pig per day from dry feeding, 7 litres from a 4:1 water–meal ratio and 14 litres from whey-fed pigs). Table 8.1 gives the NPK equivalent of *undiluted* slurry from dry meal fed and whey-fed pigs. The approximate fertilizer equivalent of 25 000 litres of average slurry from dry-fed pigs would be 500 kg of 22 : 12 : 12 fertilizer. Whey-fed pig slurry would be equivalent to only 200 kg of a similar fertilizer. Pig slurry characteristically contains a very much smaller proportion of potassium than cattle slurry and therefore gives a much smaller risk of hypomagnesaemia in sheep or cattle grazing swards to which it is applied.

As with all other farm manures, pig slurries will supply a 'bonus' of calcium, sulphur, magnesium and trace elements.

There is one most important difference between pig slurries and other manures. Because of additions to foodstuffs pig slurry contains very large amounts of copper and zinc. In extreme cases, raw pig slurry can contain as much as 60 g copper and 45 g zinc per 1 000 litres (1 tonne). Small *single* applications of some 10 000–20 000 litres/ha can thus usefully supplement the available copper and zinc already in the soil. Large, repeated applications of more than 25 000 litres/ha on a particular area can, however, increase the amounts of these elements to levels toxic to crops.

Even more important are the risks of *direct* copper and zinc toxicity to ruminant stock, especially sheep eating grass con- taminated by slurry. Pig slurry should *not* be applied to grazing land. It is best applied to bare soil for arable crops, taking care not to make repeated large dressings, and if this is not possible it should be applied to grass conservation areas either in early spring or immediately after a first silage cut to allow the slurry to be washed off the herbage.

Poultry slurry Poultry excreta are very rich in available nitrogen. Excreta produced by birds kept in battery cages on slatted or wire floors can be made into a very rich slurry in terms of NPK equivalent (Table 8.1) as can be seen by comparing it with cattle and pig slurry. In fact 25 000 litres of well-made undiluted slurry will be approximately equivalent to 1 tonne of a 24 : 12 : 12 compound fertilizer, a much higher value than cattle or pig slurry.

A thousand laying hens will produce 40–60 t of undiluted ex- creta per year. It is semi-liquid and contains approximately 25 per cent of water.

Application and use

Unfortunately, on many farms slurry is regarded as a nuisance, posing a disposal problem, rather than as a valuable source of nutrients and a small contributor to soil organic matter. The con- venience of spreading near the site of production has led to ex- cessive applications to small in-by fields giving rise to physical problems such as capping of the soil and smothering of herbage, and chemical problems such as excess nitrogen and potassium. Such applications are grossly ineffective – both losses of nitrogen by ammonia volatilization and stream pollution through leaching of nitrates can be very serious.

Also, unless adequate storage is available or if water is used too freely the farmer is obliged to spread slurry during inconvenient periods and to leave it on the soil surface with consequent nitrogen losses.

Time of application and choice of crops The most effective times to spread slurry for arable crops are in the late winter or early spring for spring-sown crops. Applications in autumn, even for autumn-sown crops, but especially for spring- sown crops, may supply phosphorus and potassium effectively but there are grave nitrogen losses. As a rough guide the fertilizer- nitrogen equivalent of 1 000 litres of cattle slurry applied in September – October for an arable crop sown the following spring is only 0.5 kg N compared with 2.5 kg N if applied in March –April. Thus slurry applied in autumn or early winter is very inefficient as a nitrogen fertilizer substitute – even for an autumn-sown crop.

Like farmyard manure, slurry is most effective for potatoes, vegetables and late-sown crops such as maize. Its use for winter barley and winter wheat is not effective.

Much slurry is applied to grassland, where it is much less effective than for arable crops. The main reason for this is the inability to incorporate the solid fraction of the slurry and the resultant nitrogen losses by volatilization of ammonia. If it is essential because of the farming system to apply slurry to grass, two applications may be made, one before growth starts and one after first silage cut.

It is important that slurry should *not* be applied to grass for grazing. Contamination of herbage may persist for weeks in dry weather and stock may well refuse to graze. The worst problem is with *pig* slurry. Herbage contaminated with this copper-rich material is highly toxic to grazing animals and especially sheep.

A further risk to animals grazing herbage treated with the potassium-rich *cattle* slurries is hypomagnesaemia. The grass takes up luxury amounts of potassium which inhibit magnesium uptake. As a result the grazing animal cannot keep up its blood magnesium levels and may die through hypomagnesaemic tetany.

Method of application and incorporation The solid fraction of separated slurries may be spread by the rotary drum machines that are used for farmyard manures. Unseparated slurries or the liquid fractions of separated slurries are pumped into mobile tankers with spray attachments for spreading the material. Nozzle blockage may be a problem with the less dilute slurries and even more so if attempts are made to use standard irrigation equipment. Only the liquid fractions of separated slurries should be used in this way.

It is essential for effective use of slurries that they should be cultivated into the surface soil immediately after application to reduce nitrogen losses.

An important aspect of slurry application in autumn and winter is the risk of damage to surface soil structure by the tracking of tankers which are very heavy when fully loaded. It is essential not to spread slurry on heavy-textured soils (clays, clay loams, silty clays) when they are wet enough to be plastic. This leads to serious surface compaction and cloddiness. Unfortunately in many north-western parts of the British Isles such soils are plastic during the whole winter period.

Rates of application Rates and frequency of application of pig slurries must be carefully restricted to avoid copper and zinc toxicity. With this proviso the rate of application of a slurry should be varied according to the estimated fertilizer nitrogen equivalent available for the first crop (Table 8.1). Thus 20 000 litres of a good cattle slurry would supply 40–60 kg of readily available nitrogen with a further 50–60 kg more slowly available.

The use of very high rates of application (more than 25 000 litres/ha) can lead to very serious contamination of grassland and may kill some of the grass. Similar applications to bare soil can leave a complete and impervious blanket of slurry capping the soil. This can lead to anaerobic conditions in the surface soil and can prevent seedling emergence.

Smell and pollution problems The smell of cattle slurry is more unpleasant to most people than that of farmyard manure and can raise objections from the local community. If well handled, rapidly spread and incorporated, the nuisance is small and brief. Poultry slurry, although not widely used, has a more objectionable smell but neither can compare with the atrocious and overpowering stench of pig slurry. This can persist for days or even weeks after application and is a real public nuisance if the slurry is applied near dwellings. Methods are available to reduce the smell to an acceptable level by vigorous aeration of the slurry, giving rise to aerobic decomposition and the destruction of the evil-smelling components. This can be followed by chemical flocculation of the slurry and the separation of relatively innocuous sludge and liquid fractions. Such methods give rise to some extra costs and are not widely used but certainly should be obligatory to avoid public nuisance.

Manures produced off the farm

Sewage sludge

'Night soil', the contents of dry closets, which preceded water closets, was used as a manure for many generations. Evidence of this is seen in the limited areas of dark-coloured soil littered with potsherds and broken glass around old dwellings in country areas. With the advent of the water closet, quite recent in many country areas, the so-called 'septic tank' was introduced near isolated dwellings, usually in an unknown and inaccessible position, providing an excellent alternately aerobic and anaerobic fermentation of human excreta. In urban areas similar but more sophisticated systems have been developed for processing sewage on a large scale. The product known as sewage sludge is made available to farmers fairly cheaply because of the need to dispose of the sludge within short haulage distances.

Several types of sewage sludge are available. Raw sewage sludges have a very objectionable smell and can contaminate vegetation or soil with pathogenic organisms such as bacteria, animal parasite eggs and potato nematode cysts. The greatest care must be exercised in the use of raw sludges. Indeed it is preferable *not* to use them at all.

Digested sludges, made by fermenting the raw material aerobically, are less objectionable in smell and pathogen content. Digested sludge is partly decomposed, contains 96–97 per cent water and can be spread directly on the soil from tankers similar to those used to distribute slurries.

Sludges, digested or raw, may be dried out to some extent by mechanical or chemical methods, to produce a solid material known as 'sludge cake' or 'dried sewage sludge'. This can be managed like farmyard manure. During water removal much of the readily available nitrogen is also taken out, so that sludge cake provides little nitrogen even for the first crop after application.

Fertilizer equivalents Sewage sludges differ from most other manures in that they contain virtually no potassium. Approximate equivalents are given in Table 8.3.

Disease hazards in the use of sewage sludge Digested sludge and even sludge cake contain disease organisms which can be transmitted through contaminated herbage or soil. Obviously risks are even greater with raw sludge.

Among the organisms that can affect human beings the most

Table 8.3   N, P₂O₅ and K₂O equivalents of some manures made off the farm.

| | Nutrients in kg/tonne | | | | | |
| | Available for first crop | | | Total | | |
	N	P_2O_5	K_2O	N	P_2O_5	K_2O
Sewage sludge:						
Raw	0.5	0.7	Nil	1.5	1.5	Nil
Liquid digested	1.7	1.0	Nil	2.0	2.0	Nil
Sludge cake	4.0	7.0	Nil	12.0	15.0	Nil
Town refuse:						
Screened dust	0.5	0.5	1.0	4.0	1.5	2.5
Compost	0.8	1.0	1.5	5.5	2.0	3.0
Seaweed	3.5	0.5	10.0	5.0	1.0	12.0

Source: Data from several sources.

important is *Salmonella* which is responsible for some gastrointestinal disorders. Because of this it is very unwise to use any kind of sewage sludge for crops to be eaten raw by human beings. As *Salmonella* can also affect grazing animals, raw sludge should never be applied to pasture. Treated sludge should be applied only if the pasture can be left ungrazed for 7–8 weeks.

There is also a small risk of the transfer of tapeworms to cattle from sewage products.

Needless to say, anyone in direct contact with sewage sludge, during or after spreading, is subject to health risks.

Toxic elements The amounts of various elements in sewage sludge can vary enormously from place to place and time to time. In country areas with little industry other than agriculture the sludges usually contain only small amounts of toxic elements. In contrast sewages containing industrial wastes can contain very large quantities of such elements. The application of such sludges to soil can lead to reduced yields. Even worse, they may not affect crop growth but may be toxic to animals or human beings eating the crops.

The main hazards to plants come from zinc, which is readily taken up, and copper and nickel, all of which can be present in large quantities in industrially contaminated sewage. Unfortunately they accumulate in the soil and if contaminated sludge is used repeatedly on a given area it may be many years before the toxic effects decline.

Boron, mainly from detergents, may be present in some sludges in sufficient quantities to be toxic to sensitive crops, especially cereals and potatoes. Unlike zinc and copper it is easily leached and does not build up long-term toxic levels.

Cadmium is the element most likely to be toxic to human beings consuming produce from sewage-treated soils, as it is readily taken up by the plant. The amounts present in different sewage sludges vary very widely, as do those of mercury, lead, selenium, chromium and molybdenum, all of which are potentially toxic to human beings in the amounts that can be taken up by plants. Fortunately lead and chromium, which are major contaminants in some areas, are not so easily taken up by the plant as cadmium.

Information on safe use is available from the Department of the Environment and the National Water Council. These authorities have recommended standards for maximum applications of toxic trace elements over a period of years. The information on which they are based is inevitably scanty. Also, because of the variability of the products, the possibility of sudden influxes of trace elements from industrial units and the difficulties of obtaining representative samples for analysis, it is virtually impossible to be sure that the empirical limits will not be exceeded if repeated applications are made at one site.

If the toxicity risks are considered along with the health hazards and the relatively low manurial value (it is consistently inferior to farmyard manure, applied at the same rate of dry matter per acre, in its effects on crop yields), any grower should give most careful consideration before using sewage sludge at all. ADAS in England and Wales and the Agricultural Colleges in Scotland will advise on all aspects of its use. My own view, reached with great reluctance because it is a useful though limited source of essential nutrients and organic matter, is that virtually any other manure is preferable, except possibly town-refuse composts.

Town refuse

Refuse disposal on the land has taken place since time immemorial. Disposal presents great problems in urban areas and as an alternative to dumping it in the sea or in old quarries some local authorities process the waste in various ways and offer the products to local growers.

In the simplest processes the waste is pulverized and screened, removing most of the broken glass and metal. The products certainly contain some organic matter, much of it paper, and a small proportion of plant nutrients. In an extensive series of field experiments by ADAS they were consistently inferior to farmyard manure applied at the same dry matter rate. One such product, 'screened dust', was on average only one-quarter as effective as farmyard manure in increasing crop yields.

Another method of treating town refuse, pioneered in Scandinavia and adopted for some time by local authorities in Scotland, involves introducing the refuse, with added water and sometimes nitrogenous materials to aid fermentation, into large revolving drums inclined a few degrees from the horizontal. The material is left in the drum for several days during which the temperature rises and some decomposition takes place. Unfortunately the raw material includes so much paper and other carbonaceous materials that the product coming out of the drum has a high carbon–nitrogen ratio and contains little phosphorus or potassium. The *total* nitrogen content can be similar to that of farmyard manure but because of the high C–N ratio there have been cases of nitrogen deficiency in crops grown after application of the material without any nitrogenous fertilizer.

The 'composts' can certainly be improved by storing in heaps for a year or so, during which there is intensive micro-organism activity, particularly of actinomycetes. Much carbon dioxide is lost and the carbon–nitrogen ratio falls as cellulose is decomposed.

Even the stored product is inferior to farmyard manure when used in the field. There is, therefore, little to recommend the use of these materials for agricultural crops and still less for market garden crops.

Seaweed

Seaweed is a valuable and quickly decomposing manure used in coastal areas but not now worth transporting for more than a kilometre or two inland. Its use, once extensive, has now been restricted by labour costs. The seaweeds most used are the brown species. *Laminaria* (tangle, kelp) grows below the low-water mark and is harvested after storms have dislodged it and deposited it on shore. *Fucus* (bladder wrack) and *Ascophyllum* (knotted wrack) grow between the high- and low-water marks and can be harvested between tides.

Seaweed is very rich in potassium (double that in farmyard manure). It also contains about the same proportion of nitrogen as does farmyard manure, very little phosphorus, but appreciable quantities of the trace elements manganese, iron and zinc. Not surprisingly it is also rich in sodium chloride.

Ideally, seaweed should be harvested, spread and ploughed in immediately, to prevent the loss of easily leached potassium. There should be no adverse effects from the sodium chloride added in this way.

Seaweed can also be composted by adding it to farmyard manure heaps. In some areas it is stored in small heaps which dry out rapidly. The nutrient equivalents of *fresh* seaweed are given in Table 8.3.

Apart from locally used seaweed there are on the market some dried seaweed meals and liquid seaweed products. They are usually expensive and are sometimes sold with extravagant claims regarding trace element content and soil-conditioning properties. In fact seaweed should be regarded as a good manure of approximately the same value as farmyard manure when applied at the same rate of dry matter per hectare.

Other bulky manures

Other organic materials are sometimes obtainable in restricted areas. They include easily decomposed wastes from breweries or distilleries, spent mushroom composts and the more slowly decomposed shredded or pulverized tree bark, sawdust and wood chips. Before using any unusual materials advice should be sought from ADAS or the Scottish Agricultural Colleges.

Crop residues as manures

All unused or unwanted crop residues left on the field after harvest have some manurial value. Normally the residues are ploughed in or otherwise incorporated and become a temporary or permanent part of soil organic matter. It is very difficult to estimate the amounts of nutrients returned to the soil in, for example, the roots and stubble of cereal crops or potato haulms but they are certainly not negligible and must be regarded as a small bonus for following crops.

There are, however, two cases worthy of more consideration – the use of cereal straw as a manure and the growing of green manure crops.

Straw as a manure

At the time of writing very large amounts of straw and cereal stubble are burned annually on the field. The fashion comes and goes but it is certainly rife in the increasing areas of intensive cereal growing and there is no doubt that many farmers regard straw as expendable and as a nuisance. Certainly when continuous winter cereals are grown the straw must be handled and disposed of rapidly during the busy autumn period. In many areas the selling price is too low to warrant transporting straw very far and non-agricultural enterprises which could use it cannot overcome the problems of bulk and transport.

Burning is encouraged by statements that weed seeds are killed and disease hazards are reduced. Both statements are to some extent true but neither should be stressed too much. There are usually sufficient disease organisms carried over when the straw and stubble are burned to inoculate a following crop and disease will spread rapidly from these centres if conditions are suitable.

There will certainly be a reduction of disease organisms such as *Rhynchosporium secalis* and *Pseudocercosporella herpotrichoides* (eyespot) after burning.

The much-made point about destruction of weed seeds by burning is less acceptable. Weed species survive by their persistence and resilience and will not be greatly reduced by burning or encouraged by straw incorporation.

Dry straw is almost all organic matter, mostly carbonaceous (carbon–nitrogen ratio: 40–80:1) and contains only small amounts of nutrients (0.4–0.5 per cent N, 0.25–0.4 per cent P_2O_5 and 0.35–0.45 per cent K_2O). If the straw is burnt there is a total loss of nitrogen and organic matter but the phosphorus and potassium are retained in the ash. Theoretically at least they will be returned to the soil but this depends on wind and rain and there is certainly no guarantee that they will remain *in situ*.

In searching for alternatives to burning, the obvious one seems to be overlooked. Straw is the normal basis of our most valuable manure, farmyard manure made with cattle excreta, which add to it nutrients and especially the available nitrogen required to help in its decomposition.

Straw incorporated directly into the soil will decompose more slowly but in fertile aerobic soils with sufficient but not excess water it can make considerable contributions to soil organic matter, especially in areas of intensive cereal production. In this way it will help to increase the water-holding and nutrient retention capacities of the soil, to improve the soil structure and to reduce insidious erosion which happens when soil organic matter levels fall below a critical point.

If continuous cereals are being grown the return of *all* straw and stubble to the soil will supply some 2.5–5 t/ha of organic dry matter *each year*, equivalent to some 10–20 t/ha of farmyard manure. This is 2–4 times as much as could be made and applied on a mixed stock/arable farm.

Figure 8.2   A straw-chopper in action. Photograph by courtesy of John Wilder (Engineering) Ltd, Wallingford, Oxon.

The management of straw incorporation is not simple and several rules must be followed for success:

● Because of the bulky nature of straw it must be chopped into short lengths (4–5 cm) before cultivating in. (Fig. 8.2). This will make incorporation easier and slightly increase the speed of decomposition.

● Because of the high carbon–nitrogen ratio of straw, extra fertilizer nitrogen may be needed when it is ploughed in. Approximately 1 unit of extra nitrogen will be needed for each 100–120 units of straw dry matter. For winter cereals this nitrogen can be applied along with some P and K as part of the autumn fertilizer programme.

● The straw may be lightly cultivated into the surface or, alternatively, ploughed in but great care must be taken to get a good spread of straw through the topsoil. This will encourage rapid decomposition. Also there is no doubt that straw which has been badly incorporated, particularly in poorly drained soil, can lie in a layer under the furrow slice and decompose anaerobically, giving rise to a noxious smell and becoming toxic to the roots of the new crop because of the production of hydrogen sulphide and other toxic substances. This problem is most likely to arise in heavy-textured soils with moderate or poor drainage in wet seasons. In such conditions ploughing-in straw is a doubtful proposition.

With this exception in mind there is no doubt that if straw is regularly returned to the soil there will develop in that soil microflora which will live on the straw and humify it. So long as conditions in the soil remain sufficiently moist and the pH is regulated by liming the only major obstacle to steady and efficient decomposition of straw is a lack of available nitrogen. The most successful approach to this problem would be to supply the necessary nitrogen as slurry immediately after chopping the straw and cultivating them in together within a few days. This would be in effect incorporating the raw materials of farmyard manure directly into the soil. Obviously such a method requiring rapid and intensive action immediately after harvest would not be feasible in all farming systems but would be most beneficial if it could be managed.

Because of the difficulties involved, adverse publicity about straw lying undecomposed for months or even years and, above all, the fact that the farmer will see little *immediate* return for his effort, there has been prejudice against returning straw to the soil. This has been strangely enhanced by the pride of the farmer and, even more, that of his ploughman in showing a neat furrow. In fact a tilth with chopped straw showing through, evenly distributed, has great advantages in surface aeration, seedling protection and, on susceptible soils, prevention of wind erosion.

For the long-term good of soil organic matter, all spare straw should be returned to the soil.

Green manure crops

Green manures are crops grown especially for incorporation into the soil without removing any part of the crop for commercial purposes. Green manuring is a very old-established process and was certainly practised by the Romans in the early centuries AD using leguminous crops such as vetches and lupins.

The essential characteristics of green manure crops are rapid growth, vigorous root development and abundant tops. Cheap, reliable and rapidly germinating seed is essential and the crop must not present difficult husbandry problems.

Green manuring crops perform several functions:

● Addition to the soil of readily mineralizable plant nutrients, particularly nitrogen but also appreciable amounts of other essential elements including the trace elements.
● Supply of organic matter to the soil.
● Provision of soil cover, important in areas where erosion is a problem.
● Prevention of leaching of nutrients by re-cycling them.
● Control of weeds by smothering them – a function which has declined in importance as weedkillers have become more effective.

Unfortunately the two main functions, supplying available nutrients and building up soil organic matter, conflict to some extent. Green manures properly used should *either* increase humus content *or* increase the immediate supply of available nitrogen and other nutrients but cannot effectively do both at the same time. Therefore a choice must be made.

If the main aim of growing a green manure crop is to increase the supply of nutrients, with emphasis on nitrogen, a legume should be grown and ploughed in while immature. This, partly because of the nitrogen-fixing symbiotic bacteria in the root nodules, gives material with a very low carbon–nitrogen ratio. After ploughing in, rapid decomposition takes place with release of available nitrogen for the following commercial crop. Much of the organic matter is lost to the atmosphere as carbon dioxide, so that the humus content of the soil is not greatly increased.

Suitable leguminous crops are red clover (*Trifolium pratense*), sweet clovers (*Melilotus* spp.), common vetch (*Vicia sativa*), trefoil (*Medicago lupulina*) and yellow lupin (*Lupinus luteus*).

If the main aim of growing the green manure crop is to maintain or increase soil organic matter it is preferable to grow a non-leguminous crop and allow it to grow further towards maturity, though still green, before ploughing in. This material will have a higher carbon–nitrogen ratio and will contribute more persistent humus to the soil. Suitable species are mustard (*Sinapis alba*), rye

(*Secale cereale*), rape (*Brassica napus*), turnip rapes (*Brassica rapa* and crosses) and ryegrasses (*Lolium* spp.).

Some growers, in the days before effective herbicides were available, were willing to sacrifice one whole season out of four or five to green manuring and cultivations. Two or three successive green manure crops could be grown during that year. The soil fertility was increased and persistent weeds were discouraged. Such practices are not regarded as economical today and green manure crops must now be fitted in between commercial crops. This can be done by sowing immediately after harvesting an early crop or by undersowing a cereal crop with a green manure crop. The latter method has become more difficult as cereal yields have increased and the undersown crop tends to be smothered. On the other hand, too vigorous an undersown crop can cause harvest problems. It might be worthwhile to use a low-growing legume such as wild white clover (*Trifolium repens*) or bird's-foot trefoil (*Lotus corniculatus*) for undersowing.

Sowing leguminous green manure crops by, for example, direct drilling into stubble after early-harvested cereals or into potato land, followed by autumn ploughing-in, can supply 40–60 kg/ha of available nitrogen for the following crop.

Fertilizers may be applied to the green crop but it is usually sufficient to rely on residual nutrients from the previous crop. Certainly no nitrogenous fertilizers should be applied to leguminous green crops.

It is important to bruise or chop bulky green manure crops thoroughly before incorporating them into the soil, in order to hasten their decomposition. Great care should be taken, as when ploughing in manures or straw, to avoid the formation of a discrete layer of organic matter, especially in soils with drainage problems.

An interesting and useful adaptation of the green manuring principle is the practice of feeding standing crops to animals in the field. This can be done on a catch-crop basis by sowing stubble turnips or rape into stubble at harvest. Stubble turnips are usually rapid growing white turnips (*Brassica campestris*, e.g. var. Debra).

They may even be sown broadcast from aircraft before harvesting the cereal to give an earlier crop. The stubble turnips may then be grazed off by sheep, the urine and faeces of which will be returned to the soil along with uneaten residues.

In some marginal areas full crops of swedes or turnips are

subjected to controlled strip grazing by sheep during autumn and winter, followed by early spring ploughing. A rich mixture of excreta and crop residues is lightly trodden in by the grazing sheep during winter when oxidative losses are at a minimum. There are management problems on heavy soils in wet winters but otherwise this is a very efficient way of recycling nutrients.

Chapter 9 Fertilizers

Fertilizers are substances which contain appreciable quantities of one or more essential plant nutrients. They are mostly inorganic. The raw materials from which they are made are atmospheric nitrogen, rock phosphates derived from large deposits in North Africa and elsewhere, mineral deposits or ores containing salts of potassium, sodium, magnesium, sulphur and trace elements.

Some of the raw materials, such as rock phosphates and limestones, may be used directly as fertilizers after being ground to a fine powder. More commonly the raw materials must be processed in some way. The obvious case is atmospheric nitrogen which is inert and useless as a fertilizer until it has been combined with carbon, hydrogen and/or oxygen to form ammonia, nitric acid, ammonium nitrate or urea. Other raw materials such as mineral deposits rich in potassium, sodium and magnesium could be, and in fact were, used directly as fertilizers but are hygroscopic and difficult to handle. This has been overcome by refining them and using conditioners to improve their physical condition.

Most modern fertilizers are designed to supply easily available nutrients in water-soluble forms for the crop grown in the year of application. There are often residual effects on crops grown in subsequent years.

Other fertilizers such as ground mineral phosphate and ground limestone are virtually insoluble in water and, when added to soil, release their nutrients gradually over a period of some years.

There are a few organic substances which, because of their high concentration of one or more nutrient elements, may be classed as fertilizers instead of manures. With the exception of urea, which is now produced synthetically on a large scale, they are usually much more expensive than comparable inorganic fertilizers and as a result are not widely used in agriculture. There is no magic about them, as is sometimes claimed and at normal rates of application their contribution towards increasing the organic matter content of soil is negligible. They include bone meal, steamed bone flour, dried blood, hoof-and-horn, guano obtained from vast deposits of bird excreta found in South America and various products from fish and meat wastes.

The purpose of fertilizer use

Fertilizers are used to supplement the nutrients which the plant can obtain from the soil in order to increase crop yields without detriment to quality. This is needed in most types of agriculture and especially in intensive systems from which high yields are sought and where the unsupplemented soil cannot supply nutrients quickly enough or in sufficient quantities to meet the requirements of crops.

Kinds of fertilizer

The farmer is offered a bewildering range of fertilizers. They may be solid and in granular form as are most fertilizers sold in the British Isles at present. They may be liquid, consisting of solutions or suspensions of fertilizer salts, usually in water. Liquid or fluid fertilizers take only a small part of the British market (about 6 per cent in 1983) but are more widely used in other countries including the USA. Gaseous ammonia is also used as a fertilizer by injecting it beneath the surface of the soil from cylinders. Again, although not widely used in the British Isles, this material takes about 50 per cent of the market for simple nitrogenous fertilizers in the USA.

In addition to variations in physical state fertilizers also vary greatly in chemical composition.

Simple or straight fertilizers are designed to supply only one nutrient element. Ammonium nitrate in the form of 'Nitram', containing 34.5 per cent of nitrogen, is a good example. Some simple fertilizers essentially used to supply one element may fortuitously provide another. For example ground mineral phosphate, essentially a phosphorus source, also contains some calcium. Similarly, ammonium sulphate, although designed to provide available nitrogen, contains rather more sulphur than nitrogen. Examples of simple fertilizers are given in Table 9.1.

Compound and blended fertilizers although made in different ways are designed to supply two or more elements. At present the great majority of these fertilizers supply only N, P and K or any combination of two of them.

Table 9.1   Some simple fertilizers and materials commonly used in compound fertilizers.

Simple fertilizers:

Nitrogen	% N
Urea	45
Ammonium sulphate	21
Prilled ammonium nitrate	34
Ammonium nitrate/calcium carbonate	21–26
Anhydrous ammonia	81
Liquid fertilizers containing ammonium nitrate, ammonia and urea	20–40

Phosphorus	% P_2O_5
Superphosphate	18–21
Triple superphosphate	45–47
Ground mineral phosphate	29–33
Basic slag	8–22

Potassium	% K_2O
Potassium chloride (muriate of potash)	60
Potassium sulphate	50

Compound fertilizers:

Nitrogen sources
   Ammonium nitrate, urea, ammonium sulphate, ammonia (in liquids only)

Nitrogen plus phosphorus sources
   Mono-ammonium phosphate (12% N, 61% P_2O_5) Di-ammonium phosphate (21% N, 53% P_2O_5)

Phosphorus sources
   Triple superphosphate

Potassium sources
   Potassium chloride, potassium sulphate

Solid fertilizers

Solid fertilizers have dominated the British market for more than a century and continue to do so. They include several widely used simple fertilizers (Table 9.1) but a large proportion are sold as 'compounds'.

Compound solid fertilizers Some modern compound fertilizers consist of granulated mixtures of substances such as ammonium nitrate (35 per cent N), triple superphosphate (45 per cent P_2O_5) and potassium chloride (60 per cent K_2O), all of which may be used as simple fertilizers. Thus a compound fertilizer consisting of these substances alone mixed in the ratio of 2 parts ammonium nitrate : one part triple superphosphate: 1 part potassium chloride would have the percentage composition in terms of N : P_2O_5 : K_2O of 17.5 : 11.25 : 15.0. These figures are obtained by dividing the percentages of N, P_2O_5 and K_2O in the constituent substances by 2, 4 and 4 respectively. (One half of the final product is ammonium nitrate with a quarter each of triple superphosphate and potassium chloride.)

The original compounds were made by mixing 'simples' together in powdered or crystalline form. They gave rise to problems of segregation of particles of different density during transport and spreading resulting in erratic distribution – one part of a field receiving much more than its share of N or P or K. The great technical achievement of granulation has solved this problem. In modern technology the final components of the fertilizer are synthesized, mixed and granulated in continuous processes giving water-soluble granules of uniform size, composition and density.

Although designed to supply N, P and K, compound fertilizers during the first half of this century also contained appreciable amounts of the other major elements – sulphur, calcium and magnesium.

Since 1945, however, there has been an increasing tendency to concentrate on N, P and K at the expense of Ca, Mg and S. This has been done by:

● Virtually eliminating magnesium from the raw potassium salts originally used in fertilizers.
● Replacing ammonium sulphate ($(NH_4)_2SO_4$), which contains 24 per cent of sulphur, as the main nitrogenous fertilizer by ammonium nitrate, (NH_4NO_3) or urea ($CO(NH_2)_2$), neither of which contains sulphur.
● Replacing superphosphate, which consists of a mixture of monocalcium phosphate ($Ca(H_2PO_4)_2$) and gypsum ($CaSO_4.2H_2O$) by ammonium phosphates, ($NH_4H_2PO_4$) and ($(NH_4)_2HPO_4$), which contain neither calcium nor sulphur.

The net result of these changes in the composition of fertilizers has been the production of a range of high-technology compound fertilizers rich in NPK. They are granular, easy to spread with modern equipment and have lost the tiresome property of absorbing water which was a serious drawback in earlier fertilizers. They also store well without coalescing in the bag.

Anyone who in the 1930s and 1940s had to cope with liquidized fertilizer running out under the door of the fertilizer store or, worse, to take a hammer to a bag of solidified fertilizer (fortunately *not* ammonium nitrate!) will appreciate much more than the younger generation the technical near perfection of the modern granular fertilizer. But it must be appreciated above all that these fertilizers supply, with few exceptions, N, P and K and completely neglect calcium, sulphur, magnesium and the trace elements.

Table 9.2   Composition of typical compound fertilizers (1940–80).

Period	% N	% P_2O_5	% K_2O	Sum
1940–1950	7	7	7	21
1950–1960	9	9	16	34
1960–1970	12	12	18	42
1970–1980	16	16	21	53

Some idea of the changes in NPK concentration over the last forty years can be gained by a comparison of typical compound fertilizers of the four decades (see Table 9.2). Thus, present-day compounds are about three times as concentrated in terms of NPK as their counterparts in 1940–50. This represents a very large saving in the cost of transporting and spreading a given amount of nutrients.

Blended solid fertilizers   In some countries, particularly the USA, 'blended' fertilizers are widely used. They consist of granules of simple fertilizers, containing N or P or K, bulk-blended to ordered ratios by mixing them, either on the farm or at a local depot. The idea is very attractive because *any* ratio of N : P : K may be formulated on the spot.

In modern blends the single components are carefully matched in terms of granule size and density in order to avoid segregation during mixing or transport. This problem does not arise with compound fertilizers.

We are unlikely to see widespread use of bulk blends in Britain during the next 20–30 years because of the commitment of vast capital investment to compound fertilizer production by the major manufacturers.

Liquid or fluid fertilizers

Most liquid fertilizers are solutions in water of the same substances that are used in water-soluble solid fertilizers. Recent advances in technology have introduced stable suspensions to which the term 'fluid' may be applied. Some materials rarely used in solid fertilizers are used in 'liquids'. They include ammonia because of its high nitrogen content (81 per cent N) and its solubility in water as well as polyphosphates, some of which are very rich in phosphorus and are soluble in water.

Liquid fertilizers are generally sprayed on to soils or 'dribbled' on to standing crops to minimize leaf scorch. Much of this work is done in Great Britain by contractors who take all risks of storage and machinery corrosion.

Liquid fertilizers were first introduced in areas such as the southern states of the USA where ambient temperatures are high and there are few problems of dissolved substances crystallizing out or precipitating. In cooler climates, including that of the British Isles, the risk is much greater, especially in winter storage.

Unfortunately once crystallization has occurred the crystals settle out, forming a discrete layer, and it is very difficult to re-dissolve them, needing both a considerable rise in temperature and vigorous stirring of the liquid over a long period. Thus storage in tanks on farms in the climates of northern Europe is difficult and it is preferable to take delivery of liquid fertilizers in bulk tankers when they are needed for spreading.

At present liquid fertilizers make up a much smaller part of the market in Britain and northern Europe (5–8 per cent) than in the USA (30–35 per cent). Most 'liquids' used in Britain are simple nitrogenous fertilizers which are much more easily handled than solutions containing N, P and K.

Farmers considering the use of liquid fertilizers should take very careful account of the advantages and disadvantages compared with solid fertilizers, especially before setting up on-farm storage.

Advantages claimed for liquid fertilizers over solids, which although small are real enough, are:

● More uniform distribution.
● Immediate absorption into soil even in dry periods.
● The possibility of incorporating some pesticide and fungicide materials.
● Lower manual labour because the liquids are handled by pumping.

● The convenience of 'deliver and spread' services.

Against these advantages must be set several possible disadvantages:

● On-farm storage problems, including corrosion of tanks and crystallization of dissolved substances.

● Risk of foliage scorch if applied to standing crops.

● The need for corrosion-resistant machinery.

● Physical damage to soil by large tanker-spreaders.

● Generally lower concentration.

It is difficult to give a fair comparison of the costs of using liquid and solid fertilizers as so much depends on the bargain struck between farmer and contractor. When comparing costs the value of immediate delivery of liquids compared with on-farm storage of solids must be taken into account.

Simple liquid fertilizers Simple liquid fertilizers are almost entirely nitrogenous. They consist of solutions of urea, ammonium nitrate and ammonia in various proportions. These simple solutions are much easier to prepare and store than compound liquid fertilizers containing phosphates and potassium salts. If ammonia is dissolved under slight pressure it is possible to attain a concentration of 30–40 per cent nitrogen in the solution.

Compound liquid fertilizers The main problem in formulating NPK liquid fertilizers is the solubility of the substances used, especially phosphates and potassium salts. In recent years progress has been made in the use of fertilizer materials suspended in the liquids with the aid of 'gelling clays' and with the introduction of polyphosphates rich in phosphorus. These developments have made it possible to increase concentrations but solid fertilizers can still be made with greater NPK concentrations than liquids. Only 3–4 per cent of the British market is taken by compound liquid fertilizers.

Gaseous fertilizers   The only commercially-used gaseous fertilizer at present is anhydrous ammonia (NH_3) (81 per cent nitrogen). It is transported as a liquid under pressure in cylinders and must be injected beneath the surface of moist soil, when it is immediately converted to ammonium hydroxide (NH_4OH).

Although not done at present it would be feasible to inject sulphur dioxide (SO_2)(50 per cent S) into the soil using similar machinery. The sulphur dioxide would be converted to sulphurous acid (H_2SO_3) then to sulphuric acid and sulphates which would supply available sulphur to the crop.

There are no gaseous potassium fertilizers and it is unlikely that any gaseous phosphorus fertilizer will be used commercially. Phosphine (PH_3) has been used experimentally. The only advantage of phosphine is its very high concentration. It contains 91 per cent of phosphorous which in our present absurd system of quoting fertilizer concentration is equivalent to 208 per cent P_2O_5. It is, however, very hazardous to use, being a deadly poison and decomposing explosively on contact with air!

Fertilizers containing nutrients other than NPK

Because of the exclusion of other nutrients from NPK fertilizers, sulphur, calcium, magnesium and the trace elements must usually be supplied separately. Some soils are sufficiently rich in these elements to need no supplementation but more and more problems are arising from magnesium, sulphur and trace element deficiencies. Fertilizers other than those supplying N, P and K are described in Chapter 12.

Fertilizer information

When considering using a fertilizer it is helpful to have information on:

● Which nutrient elements it will supply.
● How much of each nutrient it contains in percentage terms and hence the ratios of nutrients.
● The proportion of each nutrient soluble in water or, in the case of the phosphorus components, certain other solvents.
● As much information as possible on the chemical form of the nutrients.

This information will help in the selection of a type of fertilizer of a particular nutrient ratio and to calculate the amount needed to meet the requirements of a crop as decided by methods described in Chapter 14.

It is useful to have detailed information about the chemical nature of the components of a fertilizer. For example, nitrate nitrogen is easily leached from soil but ammonium or urea nitrogen is much less so. Thus a farmer working with freely drained soils, especially in wetter areas, would be well advised to use a fertilizer which contains a high proportion of ammonium or urea nitrogen.

Information on the type and solubility of phosphate in fertilizers will give a good idea of the speed of action of the material in soil. It is not, however, always easy for manufacturers to give *precise* specifications of the chemicals contained in fertilizers.

Fertilizer regulations

In the early days of fertilizer usage there were few legal controls and charlatanism was rife. Most countries have now introduced legislation which requires the manufacturer or vendor to specify enough properties of the fertilizer to define its nature.

British farmers are protected by the Fertilizer and Feeding Stuffs Acts and associated regulations which are revised from time to time. More recently the EEC has made mandatory the specification of even more details about both simple and compound fertilizers to be sold within the Community. For solid fertilizers the specifications are clearly stated on the bags. For liquid or gaseous fertilizers or for solids delivered in bulk, specifications must accompany each load. Trade names or general descriptive terms such as 'Muriate of Potash' may also be used. There are strict tolerance limits within which the actual composition of the material must fall. As an example the manufacturers of 'Nitram', a water-soluble, high-concentration simple nitrogen fertilizer are required to state on the bag:

- The fact that it is ammonium nitrate.
- The percentage of nitric (nitrate) nitrogen.
- The percentage of ammoniacal (ammonium) nitrogen.
- The percentage of total nitrogen.

Because ammonium nitrate is inflammable, the EEC regulations also require that a warning with a fire symbol should be put on the bag saying also that it 'assists fire'.

Figure 9.1 shows the panel from an EEC fertilizer bag containing prilled ammonium nitrate. The purchaser of this fertilizer should be in absolutely no doubt about its nature and composition. The regulations concerning nitrogen, potassium and magnesium fertilizers are simple and straightforward. Those about phosphate fertilizers are necessarily complex and more difficult to understand.

Nitrogen There must be a declaration of the percentage by weight (if more than 1%) of:
- Total nitrogen.
- Nitric nitrogen (nitrate).
- Ammoniacal nitrogen (ammonium).
- Ureic nitrogen (urea).
- Cyanamide nitrogen (rarely used in Britain).

Figure 9.1   *EEC fertilizer bag label showing required information on an ammonium nitrate fertilizer.*

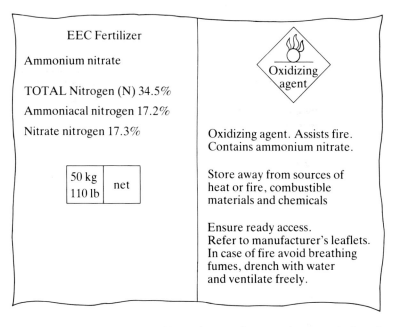

Potassium and magnesium   Potassium and magnesium are declared as percentages of the oxides K_2O and MgO soluble in water with the equivalents in terms of the elements, K and Mg, in brackets.

Phosphorus   There is a large range of phosphorus fertilizers varying from rapidly available water-soluble to ground rock phosphates insoluble in water and more slowly available to the plant. Over the years a series of extractants has been developed which will dissolve phosphates from different types of fertilizer material and which will best reflect their usefulness to plants.

Surprisingly the EEC has got agreement on the best extractants to use. There are six of them. The regulations require a statement of the nature of the phosphate. The vendor must also specify, in terms of phosphorus pentoxide (P_2O_5) with the equivalent as the element P in brackets, of:

● Total percentage.
● Percentage soluble in the specific solvent for that type of phosphate fertilizer.

Table 9.3   EEC requirements for assessment of phosphorus fertilizers.

Type of fertilizer	Statement required
Superphosphate, triple superphosphate	Per cent soluble in water. Per cent soluble in neutral ammonium citrate.
Basic slag	Per cent soluble in mineral acids (total). Per cent soluble in 2% citric acid. (At least 75% of the *total* P_2O_5 content must be citric-soluble. Only material containing 12% or more of P_2O_5 may be designated as an EEC fertilizer.)
Rock phosphate (ground mineral phosphate). Mixtures of basic slag and rock phosphate	Per cent soluble in mineral acids (total). Per cent soluble in 2% formic acid.
Aluminium-calcium phosphate	Per cent soluble in mineral acids (total). This must be more than 30% P_2O_5. Per cent soluble in alkaline ammonium citrate. This must be more than 75% of the total.

Table 9.3 gives the EEC requirements for some main types of phosphate fertilizer.

At first inspection the requirements set out in the table may seem unnecessarily complicated. Actually they are as simple as could be devised to define the products. The percentages of P_2O_5 soluble in the selected extractants will give an excellent assessment of the material in its class. For example, the greater the proportion of the total P_2O_5 of rock phosphate that is soluble in 2 per cent formic acid, the more effective it will be as a phosphate source for the plant; the term 'Soft ground rock phosphate' may be applied only if the rock contains more than 25 per cent P_2O_5 of which at least 55 per cent (about 14% of the whole material) is soluble in 2 per cent formic acid.

Compound fertilizers   The term 'compound' fertilizer is not used in the EEC regulations but is widely used and accepted in the British Isles. The EEC designate these fertilizers as NPK, NK, NP or PK according to their constituents. To qualify as an EEC fertilizer these materials must contain at least 3 per cent N, 5 per cent P_2O_5 and 5 per cent K_2O. Also the sum of percentages of N, P_2O_5 and K_2O, sometimes known ludicrously as 'total plant food', must be more than 20 per cent if all three nutrients are present or 18 per cent if only two (NP, NK, PK) are present. These restrictions present few problems to the British manufacturer because of the high concentration of our fertilizers.

The same tests and standards are applied to compound as to simple fertilizers. If the phosphates included in a compound fertilizer are not water soluble this must be implied in the name of the fertilizer – for example, 'NP fertilizer containing soft ground rock phosphate'.

Figure 9.2 shows a typical declaration on a bag of NPK fertilizer according to EEC regulations. By tradition and international agreement, if a fertilizer is labelled as, for example, 13 : 13 : 20, it is understood that the first figure refers to per cent N, the second to per cent P_2O_5 and the third to per cent K_2O.

Figure 9.2   Information on a bag of EEC NPK fertilizer.

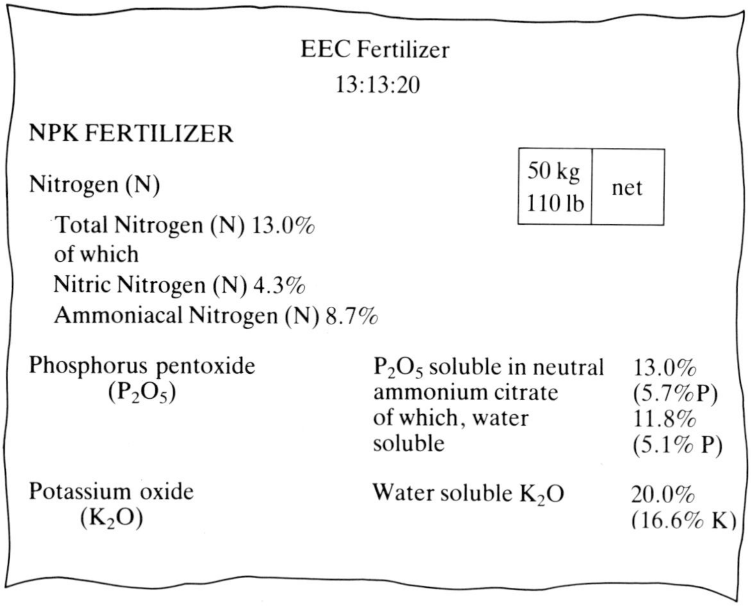

EEC Fertilizer
13:13:20

NPK FERTILIZER

Nitrogen (N)

Total Nitrogen (N) 13.0%
of which
Nitric Nitrogen (N) 4.3%
Ammoniacal Nitrogen (N) 8.7%

50 kg
110 lb   net

Phosphorus pentoxide (P_2O_5)	P_2O_5 soluble in neutral ammonium citrate	13.0% (5.7%P)
	of which, water soluble	11.8% (5.1% P)
Potassium oxide (K_2O)	Water soluble K_2O	20.0% (16.6% K)

Percentage composition of fertilizers

The amount of nitrogen in a fertilizer is always expressed as per cent N. It would be logical to express phosphorus in terms of per cent P and potassium as per cent K. Most countries, including the United Kingdom, for legal and commercial reasons have retained the very old established system of expressing the phosphorus and potassium contents as P_2O_5 and K_2O respectively. The historical reasons for this, dating back at least 100 years, are the traditional practice of analytical chemists of reporting results of ash analyses in terms of the oxides of the elements, e.g. CaO, MgO, K_2O. Partly because nitrogen compounds are burnt off during ashing for analysis and partly because nitrogen forms a range of oxides (e.g. N_2O, NO, NO_2), any of which could be used to express the nitrogen content of a fertilizer, the analysts arbitrarily settled upon using percentage N. As a result the use of N : P_2O_5 : K_2O became part of early UK fertilizer legislation. The EEC regulations give tacit approval to the system by using it for their main specifications and including P and K equivalents only in parentheses.

The system is inherently absurd. No fertilizers actually *contain* P_2O_5 or K_2O and it is theoretically possible under the system, as with phosphine (PH_3), for a fertilizer to 'contain' more than 208 per cent P_2O_5! There is, however, no pressure from manufacturers and vendors to change it, perhaps because 60 per cent K_2O looks rather better on a fertilizer bag than 53.6 per cent K, although they are in fact equivalent (1 part of K = 1.12 parts K_2O). Even more so, 23 per cent P_2O_5 gives a much better impression than a mere 10 per cent P (1 part of P = 2.3 parts of P_2O_5).

The system is unlikely to change and has, therefore, been used reluctantly throughout this book.

The percentage composition of a fertilizer is determined by the types and amounts of the chemical substances it contains. In early fertilizers a fairly large proportion of filler or conditioner such as china clay containing no N, P or K was added, thereby reducing the NPK concentration. Nowadays very little conditioning material is used and nearly all the contents of a fertilizer are active. An excellent example of this is the simple nitrogenous fertilizer 'Nitram', the active ingredient of which is ammonium nitrate. The theoretical percentage of nitrogen in ammonium nitrate is 35. Nitram is guaranteed at 34.5 per cent N. This represents a remarkable technical achievement in rendering a potentially inflammable, explosive and, in moist conditions, hygroscopic material safe and easy to use with the sacrifice of only 0.5 per cent nitrogen. In other words, in the case of Nitram less than 1.5 per cent of the total weight is 'conditioner' and not ammonium nitrate. The concentration of nitrogen in ammonium nitrate is otherwise restricted only by the presence of other elements – in this case, NH_4NO_3, hydrogen and oxygen which are an essential part of its constitution.

Figure 9.3 shows the percentage composition of some simple and compound fertilizers. For those fertilizers containing phosphorus and potassium the P_2O_5 and K_2O 'contents' are given alongside the true composition of the material. Note the very small amount of filler used in each case.

Because of the nature of the substances involved and their necessary content of non-nutrient elements it is very rare that a

Figure 9.3 Percentage composition of some simple and compound fertilizers.

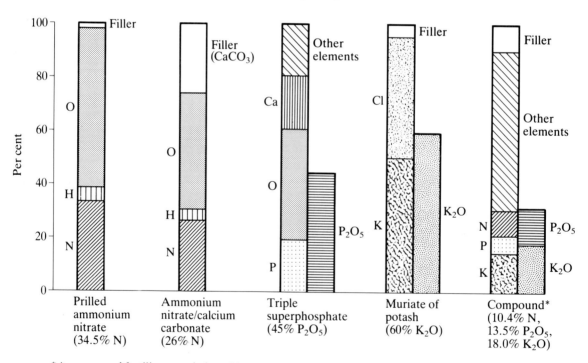

*A compound fertilizer consisting of 3 parts (by weight) of ammonium nitrate, 3 parts triple superphosphate, 3 parts muriate of potash and 1 part of filler.

fertilizer contains anything approaching 100 per cent of essential nutrients. Elemental sulphur (virtually 100 per cent S) can be used as a fertilizer. Ammonia gas, if pure, contains approximately 82 per cent of nitrogen. Normally solid or liquid fertilizers whether simple or compound contain no more than 50–55 per cent of N + P_2O_5 + K_2O.

Nutrient ratios

The nutrient ratio of a compound fertilizer can be calculated, in the simplest cases, by dividing the percentage of each nutrient by the percentage of that present in the smallest proportion. Thus a 24 : 12 : 12 fertilizer has a nutrient ratio of 2 : 1 : 1; a 12 : 12 : 18 fertilizer, 1 : 1 : 1.5. Indeed compound fertilizers are commonly formulated with fairly simple nutrient ratios although there is no rational basis for this.

Manufacturers tend to produce a series of fertilizers with set ratios and offer them for specific purposes. A compound fertilizer with the analysis 25 per cent N, 10 per cent P_2O_5, 10 per cent K_2O (nutrient ratio 2.5 : 1 : 1) might be labelled 'Cereal Fertilizer' because cereal crops generally require a much higher input of nitrogen than of phosphorus or potassium. A 12 : 12 : 18 fertilizer (nutrient ratio 1 : 1 : 1.5) might be dubbed 'Potato Fertilizer'. This certainly eases the way for the farmer and generally these fertilizers will be suitable for the stated crops but such labelling should be used only as a very general guide to the ratio of nutrients needed for the stated crop in average conditions. Compared with the 'Potato Fertilizer' quoted above it would be reasonable, on a high phosphate soil, to select a 1 : 0.5 : 1.5 fertilizer while for a phosphate deficient soil a 1 : 1.5 : 1.5 or even a 1 : 2 : 1.5 might be needed.

It is an amusing sidelight on nutrient ratios that fertilizers labelled 'General Fertilizer' or 'All-purpose Fertilizer' have commonly a 1 : 1 : 1 ratio. The implication seems to be that with a 1 : 1 : 1 ratio you cannot go wrong! This is absurd, especially if one considers that by stating N : P : K instead of N : P_2O_5 : K_2O the ratio would become 1 : 0.44 : 0.83. There will certainly be many occasions on which a 1 : 1 : 1 fertilizer would be a reasonable selection, but the term 'General Fertilizer' is best ignored.

The range of nutrient ratios in present-day fertilizers is very wide indeed and, taken along with the claims made by vendors for their own formulations, can be confusing. In fact minor variations in ratios are seldom of any practical significance because the requirements of crops for N, P and K cannot be predicted sufficiently accurately. At the time of writing there are three 'potato fertilizers' on the market with ratios of 1 : 1 : 1.4, 1 : 1 : 1.5 and 1 : 1 : 1.6. There is no effective difference between them.

Chapter 10 The main N, P and K fertilizers

The most commonly used fertilizers vary from country to country. Urea is very widely used in India and Italy whereas ammonium nitrate is the most common nitrogenous fertilizer in the United Kingdom and the Federal Republic of Germany. Dicalcium phosphate is little used in Britain but is used much more in other EEC countries. None the less there are several fertilizer materials that are of major importance. Other materials, widely used in the past, have been largely superseded. The prime example of this is superphosphate which dominated the British market for more than a century between 1850 and 1960 and is now difficult to obtain.

Table 9.1 (p. 111) gives a list of the most commonly used N, P and K simple fertilizers and the materials most used in compound fertilizers.

Nitrogenous fertilizers

Ammonium nitrate

In the British Isles a concentrated form of ammonium nitrate is manufactured and sold in the form of prilled granules guaranteed to contain 34.5 per cent nitrogen. The prilling process consists essentially of coating the granules to protect them to some extent against absorbing moisture in humid conditions and greatly to reduce the risk of fire or explosion when handling the dry material. None the less, when the prills are added to soil the ammonium nitrate dissolves quickly and becomes available within hours unless conditions are excessively dry.

Reasonable care is required in handling and storing this fertilizer. Warnings are printed on the bag that it is an oxidizing agent and 'assists fire'. It is safe if stored away from sources of heat or fire, in tough polythene bags, avoiding such foolish acts as walking over the stacks in hobnailed boots. It is *imperative* not to store it loose or in burst bags where it might come into contact with wood shavings, sawdust, straw or other combustible materials. Once spread on the soil there is no risk of conflagration.

Because of its tendency to absorb water, even in the prilled state this fertilizer should not be left in the distributor longer than is necessary to spread it.

In common with other fertilizers containing ammonium compounds, concentrated ammonium nitrate will acidify the soil and extra lime will be required to neutralize this acidity except in naturally calcareous soils.

Both ammonium and nitrate ions are available to plants. Because of this and the solubility of ammonium nitrate in water it is readily absorbed by plant roots. Unfortunately the nitrate component, half of the total nitrogen, is easily leached. There are cases of the loss of *all* nitrate from the top 25 cm of soil during short periods of heavy rain and in wet springs substantial losses will occur. For this reason the time of application of this fertilizer is critical and applications to bare ground should be delayed as long as possible. The ammonium ion acts as an exchangeable cation and is retained by the cation-exchange complex until converted to nitrate by bacteria.

Ammonium nitrate is ideal for top-dressing a growing crop, especially grass, because the established root system can catch the nitrate as it is leached downwards. There is some risk of loss of nitrogen by volatilization of ammonia if it is applied to the surface of naturally calcareous or newly limed soils. In acidic soils the risk is negligible.

Ammonium nitrate-calcium carbonate mixtures

These mixtures are now classified in the EEC Regulations as 'calcium ammonium nitrate', a misnomer. This type of fertilizer was first produced by ICI more than half a century ago under the trade name 'Nitro-chalk'. The admixture of finely divided calcium carbonate serves a similar purpose to prilling – reducing absorption of water in humid conditions and cutting down the fire risk in dry conditions. Ammonium nitrate-calcium carbonate mixtures contain much less nitrogen than the prilled form resulting in higher transport and spreading costs, but they are also less acidifying because of the built-in neutralizing effect of the calcium carbonate component.

Table 10.1 Composition of ammonium nitrate/calcium carbonate fertilizers.

N (%)	Ammonium nitrate (%)	Calcium carbonate (%)
15	42.8	57.2
21	60.0	40.0
26	74.3	25.7
34.5	98.5	Nil

Table 10.1 compares the proportions of ammonium nitrate and calcium carbonate in the various solid ammonium nitrate fertilizers.

The original 'Nitro-chalk' contained only 15 per cent nitrogen and had little or no acidifying effect. Its successors contain more ammonium nitrate and hence less calcium carbonate and cause some acidification, but not so much as the prilled concentrated form. The more recent formulations are as effective per unit of nitrogen as the original form but should be regarded as mildly acidifying – some 25–35 per cent as acidifying as pure ammonium nitrate.

Urea

Urea is an organic fertilizer, being present in the urine of mammals as the chief form of nitrogenous waste material from the body. It has now been produced synthetically for many years and it is this material that is used as a fertilizer. It is cheap to produce, especially where there is ample electric power, and in many countries is used more extensively than in the British Isles. Urea is the most concentrated nitrogenous fertilizer available as a solid and is easily soluble in water. Its formula is $CO(NH_2)_2$ and, if pure, it contains 46.6 per cent of nitrogen. Pure urea is hygroscopic and difficult to use but fertilizer technology has overcome this by prilling with a small amount of conditioner to give a commercial product containing 45 per cent nitrogen.

When applied to moist soil, urea dissolves and is rapidly converted to ammonium carbonate which is alkaline. Thus, if applied as a top dressing urea can raise the pH of the surface centimetre or so of the soil for a period of a few days. This temporary alkalinity gives rise to some risk of loss of ammonia by volatilization but the risk is even greater on chalk or limestone soils which are naturally alkaline.

There is evidence that large dressings of urea in the vicinity of germinating seeds or young seedlings as in combine drilling can damage them and also reduce or delay emergence. It was originally thought that the cause was an impurity in synthetic urea called biuret which is toxic to plants but although the amount of this substance in commercial urea is strictly controlled by EEC regulations to very low levels the ill effects have continued. They may be caused by release of ammonia within the soil. Whatever the cause, it is wise to cultivate newly applied urea well into the top 10–15 cm of surface soil.

Following the brief period of alkalinity caused by its conversion to ammonium carbonate or ammonia, urea nitrogen acts in essentially the same way as any ammonium compound. It is absorbed as an exchangeable cation and is therefore not easily leached. After conversion to ammonium, urea has a strong acidifying effect on the soil which can be greater than that of ammonium nitrate, and this necessitates the use of extra lime.

There have been many experimental comparisons of the effects of ammonium nitrate and urea on crop yields. Some results favour one, some the other, with perhaps a slight overall advantage to ammonium nitrate. Urea tends to be inferior as a top dressing in dry areas of calcareous soil, but should be safe to use as a top dressing in wetter areas of acidic soils. Urea may be superior for arable crops in wetter areas of non-calcareous soils especially if the fertilizers are well incorporated. This is almost certainly due to preferential leaching of the nitrate fraction of ammonium nitrate.

Ammonia   Liquefied anhydrous ammonia (NH_3)(82 per cent N) is widely used as a fertilizer in the USA but not so extensively in Europe. Ammonia is a very toxic gas. It can be produced practically free from impurities and as such it is the most concentrated nitrogenous fertilizer. Its gaseous nature and poisonous properties, however, present difficulties.

It was pioneered as a fertilizer in the vast areas of uniform stone-free silty wind-borne soils of the USA. Injection is made 10 cm or more below the surface of the soil in a slit made by a coulter. The surface of the slit must be closed immediately afterwards. This is easily done in such soils and, for example, in the stone-free silts and peats of the East Anglian fen. Unfortunately many other British soils, particularly in northerly glaciated areas, are stony and variable in texture and structure. As a result major problems are found in using the material.

Liquefied anhydrous ammonia is a remarkable example of the difficulties of transferring a technique developed successfully in one set of conditions to an entirely different set of conditions. The problems are worthy of detailed consideration.

In stony or poorly structured soils the closing of the slit after injection is virtually impossible and serious losses of ammonia can occur.

Because of the draught required to pull the coulters through heavy soils the space between slits needs to be about 50 cm. In soils with a high cation-exchange capacity ammonia not lost immediately by volatilization is absorbed as the ammonium cation and does not diffuse adequately into the zones between the slits. The results of this can be best seen after application to grassland. There is a good yield response around the slits and poor in between, giving a 'wavy' appearance.

Immediately after application the strongly alkaline ammonia increases the pH of the soil near the slits well above the normal soil range. Also, in that zone the concentration of ammonia can be so high that the soil is virtually sterilised – young plant roots, worms and other soil organisms being killed.

Injection of large dressings, up to 300 kg N/ha into grassland in springtime, has been recommended by some operators on the grounds that one application will then be sufficient for the season. It is tempting to follow this recommendation, especially as the unit price of liquefied anhydrous ammonia can fall dramatically at high application rates. The reason for this is that the price includes the cost of application by a skilled operator. The ammonia itself is cheap and it is as easy to inject a large amount as a small amount. Experimentation has shown, however, that large applications (200–300 kg N/ha) are either taken up early in the season and thus, in a cutting system, removed in early cuts of grass, or lost from the soil. As a result the effects of such applications do not last beyond midsummer.

It has also been claimed that injections can be made in autumn or early winter for arable crops to be grown in the following season. This may well be so in the Corn Belt of the USA where soils are frozen during the winter and leaching losses small. There would, however, inevitably be leaching losses in British conditions, especially in mild wet autumns and winters during which some of the ammonia would be converted to nitrate. If used for arable crops in the British Isles, injection should be made in early spring.

In common with urea, following its transient effect in raising soil pH, anhydrous ammonia acidifies the soil and extra lime will be needed to combat this.

Because of its extreme toxicity to human beings, the storage, transport and application of liquefied anhydrous ammonia should be in the hands of experienced operators, with all the necessary safety precautions. Storage of cylinders of liquefied anhydrous ammonia on the farm should be avoided. Although normally perfectly safe, tampering with them could lead to tragedy.

Liquid nitrogenous fertilizers

Like liquefied anhydrous ammonia, these fertilizers have been included as a 'main' fertilizer because of their widespread use in the USA.

Aqueous ammonia solutions are made by simply dissolving gaseous ammonia in water. If dissolved under slight pressure, solutions containing 20–30 per cent nitrogen can be made. Although they should be placed below the surface of the soil to avoid ammonia loss they are easier to handle than liquefied anhydrous ammonia, there being less risk to operators and less loss of ammonia in difficult soils.

Other nitrogenous solutions are made using urea or ammonium nitrate or mixtures of the two. Solutions containing 20–30 per cent N can be made, transported to the farm in tankers and spread directly. The nitrogen content of such solutions may be increased to 30–40 per cent N by dissolving, under slight pressure, some gaseous ammonia, thus creating a concentrated liquid fertilizer to compete with solids of similar concentration.

All such solutions may be applied by spraying on to bare soils and incorporating immediately a day or two before sowing arable crops. This has the advantage of giving very even distribution. Problems of losses by volatilization of ammonia will be similar to those for solid fertilizers.

If sprayed on to growing crops there is a considerable risk of scorching the leaves. Crops are said to recover quickly from such scorch but some loss of crop is inevitable. Damage can be reduced in grassland and more so in row crops by the use of large droplet sprays which do not adhere easily to the leaves or by replacing the sprayer with a boom with pipes spaced at 12–18 cm pointing downwards and ending in plastic tubes which reach the soil surface. Only vegetation immediately below the tubes is at risk from scorch. This method does not give the even distribution achieved by spraying.

Phosphorus fertilizers

Most of the phosphorus, almost 90 per cent, applied in UK fertilizers is incorporated in compound fertilizers. There are two main simple phosphate fertilizers: triple superphosphate and ground mineral phosphate. Almost all modern phosphorus fertilizers are derived directly or indirectly from mineral phosphates found in North Africa, Florida, USSR and many other parts of the world. Phosphate rocks vary in composition from source to source but invariably the phosphate is in the form of tricalcium phosphate ($Ca_3(PO_4)_2$), usually in combination with calcium fluoride or carbonate. These forms are collectively known as apatite. The P_2O_5 content of a good quality rock phosphate will be around 30 per cent.

Ground mineral phosphate (GMP)

This fertilizer is made, as its name suggests, by simply grinding the rock to a fine powder so that about 90 per cent of it passes through a 100 mesh sieve (apertures approximately 0.015 mm). GMP is most effective in acid soils of pH less than 6.0 in the wetter, northerly and westerly parts of the British Isles. It is insoluble in water and becomes available to the plant only through the action of soil acids and organisms and is fundamentally a long-term fertilizer. The fine grinding provides a large surface area in order to assist this action. Some producers have marketed an ultra-fine flour-like material, much of which passes a 300 mesh sieve, but this is generally little more effective than ordinary GMP in increasing phosphorus uptake by plants.

Softness is an important property of mineral phosphates. GMP from hard rocks acts much more slowly and less effectively than that from soft rocks. Most of the GMP used in the British Isles is made from soft North African rock. The softness and effectiveness of various samples of GMP can be compared by treating with 2 per cent formic acid which will dissolve a greater proportion of the phosphate from soft materials than from hard ones. In the EEC regulations a ground mineral phosphate must have 55 per cent of its total P_2O_5 content soluble in 2 per cent formic acid in order to be described as 'soft'.

Many comparisons have been made in field experiments of the effects of GMP with those of water-soluble phosphates on the yield and phosphorus uptake of crops. Early field experiments, especially on arable crops, were based on a uniform time of application, immediately before sowing the crop. This put the more slowly available GMP at a strong disadvantage. More recently work by ADAS has shown that a soft GMP applied in the previous autumn can be equivalent unit for unit with water-soluble phosphates applied in the following spring. Most of this work was done on acidic soils in high rainfall areas.

Mineral phosphates are least effective in near neutral or alkaline soils (pH greater than 6.5) and should certainly not be used on the dry soils in south-east England derived from chalk or limestone.

They are best used in wetter areas (rainfall more than 900 mm per year) and are most suitable for upland or marginal grassland or for crops such as kale, rape or swede turnips.

Ground mineral phosphates are not pleasant to handle because of their powdery nature. Because of this a GMP has been produced in the form of granules which disintegrate rapidly in the soil.

Triple superphosphate

This water-soluble fertilizer is essentially mono-calcium phosphate $(Ca(H_2PO_4)_2)$ approximately 45 per cent P_2O_5. It is made by treating rock phosphate with sulphuric acid to give ortho-phosphoric acid (H_3PO_4) and then reacting this with more rock phosphate. Because of its water solubility the phosphorus in triple superphosphate is easily available to crops for several weeks after application. It is progressively subject to fixation as the season goes on and supplies only small amounts of residual phosphorus for subsequent crops.

Potassium fertilizers

Only very small amounts of simple potassium fertilizers are used in the British Isles. More than 95 per cent of the total amount used is included in compound fertilizers. In the USA, however, more than half of a very large annual tonnage is applied 'straight'. There are two main potassium fertilizers, potassium chloride and potassium sulphate.

Potassium chloride

Potassium chloride, KCl, is usually sold as muriate of potash containing 60 per cent K_2O, all of which is soluble in water. The term muriate is archaic, dating from the time when hydrochloric acid was called muriatic acid. It is the commonest potassium fertilizer in most countries, made from raw potash salts first extracted from mineral deposits at Stassfurt in Germany. Deposits are now exploited in north-east England, Israel, USSR and many other countries.

It is now marketed both as a dry free-running powder and in the form of granules.

Potassium sulphate

Potassium sulphate (K_2SO_4) 50 per cent water-soluble K_2O, 18 per cent S, is used to a limited extent as a potassium fertilizer for potatoes, vegetables and fruit crops in which it enhances the quality as compared with potassium chloride. It is more expensive than the chloride form but can be used to advantage in, for example, seed potato production, increasing the proportion of seed-sized tubers in the crop and resulting in higher dry matter in tubers and better keeping quality. Claims are also made that potassium sulphate, applied at equivalent rates to the chloride, gives better flavoured fruit and vegetable crops.

Potassium sulphate is used specially in countries with saline soils because the chloride form would add to the salinity problem.

It is a useful source of available sulphur.

The main constituents of NPK fertilizers

Because of modern methods of manufacture in which some components of compound fertilizers are synthesized and simultaneously mixed with others it is not always possible to identify precisely some individual substances, particularly phosphates, in the final product. The manufacturer is not required to specify the actual phosphates contained in water soluble form but simply the percentage of water-soluble P_2O_5.

Table 9.1 gives a list of the main components of compound fertilizers. The only ones not previously described are mono- and di-ammonium phosphate.

Mono-ammonium phosphate ($NH_4H_2PO_4$) 12 per cent N, 61 per cent P_2O_5 and *di-ammonium phosphate*, (($NH_4)_2HPO_4$) 21 per cent N, 53 per cent P_2O_5 are made by reacting ammonia with phosphoric acid. They are usually synthesized during the manufacture of compound fertilizers. They are both water soluble and because of their high P_2O_5 and smaller N contents they are very useful to the manufacturer of concentrated compound fertilizers. Their phosphorus component is similar in action to triple superphosphate and their ammonium component behaves similarly to that in ammonium nitrate.

Chapter 11 Other NPK fertilizers

Those fertilizers described in Chapter 10 make up a very large proportion of the present-day market. There are many other less widely used N, P or K fertilizers. Some were once widely used and have been superseded. Others have not yet found favour because of problems of manufacture, price or scarcity.

Nitrogenous fertilizers

Ammonium sulphate ($(NH_4)_2SO_4$) 21 per cent N, 24 per cent S, known in commerce as sulphate of ammonia, dominated the market as a water-soluble nitrogenous fertilizer in the first half of this century. It has gradually been replaced during the last twenty years by ammonium nitrate and urea. In its day it was the spearhead, along with superphosphate, of the great yield increases produced by fertilizers between 1900 and 1950.

It is water soluble, not easily leached because its nitrogen is in the ammonium form, but acidifies the soil very strongly. It is a very useful source of available sulphur.

Calcium nitrate (so-called), approximately 15 per cent N, is actually a complex double salt of calcium and ammonium nitrates. It is sold in prilled form, is readily available to the plant but easily leached. It is little used in the British Isles.

Calcium cyanamide ($CaCN_2$) 21 per cent N, was originally used in Britain as a combined weedkiller and nitrogen fertilizer before the Second World War. It is now used mainly in Germany and Japan.

Because of its immediate herbicidal effect it should never be applied to a standing crop or immediately before sowing. To get maximum weed killing and minimum crop damage it should be applied 2–3 weeks before sowing a crop, during which time it is converted in the soil first to urea and then to ammonium carbonate.

Sodium nitrate ($NaNO_3$) 16 per cent N, is a fast-acting, easily leached material which was one of the first nitrogenous fertilizers used. It was imported in large quantities from Chile in the nineteenth century but is expensive and is now little used in the British Isles.

Slow-release nitrogenous fertilizers

In order to reduce leaching losses and in an attempt to release available nitrogen into the soil when it is most needed by crops, a range of slow-release fertilizers is developing. They are at present very expensive and are not widely used in agriculture. If they could be made cheaply and designed to release available nitrogen to order, they might become important in the future.

Phosphorus fertilizers

Superphosphate is a mixture of mono-calcium phosphate ($Ca(H_2PO_4)_2$) and gypsum ($CaSO_4.2H_2O$), usually containing 18–21 per cent P_2O_5, mainly water soluble, and 10–12 per cent sulphur. This fertilizer was the outcome of the first successful attempt to convert insoluble bone or mineral phosphates to water-soluble form. This was done by treating with sulphuric acid. Superphosphate, a superb fertilizer, dominated both the simple and compound phosphate fertilizer market for 100 years between 1850 and 1950. It has now been superseded by more concentrated water-soluble phosphates and is little used in the British Isles.

Basic slag is, or was, a by-product of steel manufacturing processes now largely superseded. It contains both lime and phosphates. Like GMP it is ground to a very fine powder before use. It is a very dense material and is objectionable to spread in its powdered form. Because of this 'mini-granules' have been produced which are much less dirty and offensive. In the heyday of the British steel industry basic slag was the outstanding 'simple' phosphorus fertilizer and was very widely used on grassland. There is now little or no home-produced basic slag although some is imported from the USSR and the continent of Europe.

Basic slag is very variable in composition. The total P_2O_5 content can vary from 7 per cent to more than 20 per cent, the lime content from 50 to more than 70 per cent $CaCO_3$. There are also useful impurities including magnesium as well as trace elements such as manganese and iron.

Quotations are required by the EEC of the percentage of the total P_2O_5 content which is soluble in 2 per cent citric acid. This is related to the effectiveness of the slag in increasing phosphorus uptake by the plant.

The best approach to purchasing basic slag is to seek a slag with as high a P_2O_5 content as possible, at least 80 per cent of which is soluble in 2 per cent citric acid. It is best used on hill or marginal grassland but may also be used for arable crops on both acidic and neutral soils. Like GMP it is fundamentally a long-term fertilizer.

Ammoniated superphosphate is the product of treating superphosphate with ammonia. This introduces some available nitrogen and reduces the proportion of water-soluble phosphate by conversion to di-calcium phosphate. The theory behind this is to supply a mixture of immediate-acting and long-term phosphorus sources.

Di-calcium phosphate $(Ca_2(HPO_4)_2)$ 40 per cent P_2O_5, is chemically intermediate between the water-soluble mono-calcium phosphate in triple superphosphate and the insoluble tri-calcium phosphate which occurs in rock phosphates. It is only slightly soluble in water but is tested under EEC regulations by its solubility in alkaline ammonium citrate. Although it is an effective simple phosphate fertilizer it has never become popular in the British market. One reason for this is its powdered form which makes it difficult to apply. Another is that despite its effectiveness it is not water-soluble. Until recent years substantial government subsidies were given to water-soluble phosphates and this established a tradition of using such materials preferentially.

Nitrophosphates are made by treating rock phosphates with nitric acid instead of the sulphuric acid used in the production of most water-soluble phosphates. By adjusting the amounts of nitric acid used the proportion of water-soluble to insoluble phosphate can be varied. Imported nitrophosphates are on sale in the British Isles. When considering purchase it is necessary to take account of the proportion of the total P_2O_5 which is water soluble and to compare in terms of unit costs with other phosphate sources.

Condensed phosphates are a wide range of phosphates which have not yet found a significant place in fertilizers. This they may well do in the future, because of their high P_2O_5 content (55–75 per cent). Some of them are soluble in water and have been used in liquid compound fertilizers. They seem to be easily available to crops.

Potassium fertilizers

Potassium nitrate (KNO_3) 13 per cent N, 44 per cent K_2O, takes only a small share of the market. It is very soluble, rapidly available but easily leached. *Chilean potash nitrate* is a cruder product containing only 10–15 per cent K_2O and 9–18 per cent of sodium.

Sulphate of potash magnesium, 28 per cent K_2O, 11 per cent MgO, is a water-soluble double salt used in parts of Europe.

Rock potash is produced by grinding rocks containing potassium-rich silicate minerals, mainly micas. It is used to increase the reserves of weatherable potassium in the soil. It has been used in Scandinavia but not yet in Britain. There are potential sources in schistose rocks in Scotland.

Chapter 12 Fertilizers other than NPK

Because of the concentration of compound fertilizers on N, P and K the other major elements, calcium, sulphur and magnesium, as well as the trace elements need to be supplied separately. At present this is often neglected but as the stress on these available nutrients from the soil has been accentuated by many modern farming practices the need for supplementation has become steadily greater.

A list of fertilizers supplying nutrients other than NPK is given in Table 12.1

Calcium fertilizers *Calcium carbonate* ($CaCO_3$) in the form of chalk and limestone is the only regularly used specific calcium fertilizer. Chalk and calciferous limestones contain 35–38 per cent Ca and magnesian limestones contain 20–34 per cent Ca. They are insoluble in water and their calcium is slowly available. Regular liming will ensure that there is no calcium deficiency in field crops. Other fertilizers which supply some calcium are ammonium nitrate-calcium carbonate, 10–16 per cent Ca; basic slag, 15–20 per cent Ca; triple superphosphate, 16–21 per cent Ca; ground mineral phosphate, 35–40 per cent Ca; and gypsum, 23 per cent Ca.

Magnesium fertilizers *Magnesian limestone*, 5–20 per cent MgO, consists mainly of calcium carbonate and magnesian carbonate in varying proportions. A good sample would contain about 40 per cent of $MgCO_3$ (20 per cent MgO). It is by far the cheapest source of magnesium and may be used on acidic soils as part of the liming programme. It should not be used on calcareous soils because it would enhance the risk of trace element deficiencies.

Calcined magnesite, 90 per cent MgO, is a useful concentrated but slowly available source of magnesium. Because of its high concentration it could easily be incorporated in compound

Table 12.1   Some fertilizers supplying nutrients other than NPK.

Calcium	% Ca
Ground calciferous limestone (90% $CaCO_3$)	36
Ground magnesian limestone (50% $CaCO_3$)	20
Basic slag	15–20
Gypsum	23
Magnesium	% MgO
Ground magnesian limestone (40% $MgCO_3$)	16
Calcined magnesite	90–95
Kieserite	29
Epsom salt	16
Sulphur	% S
Ammonium sulphate	24
Gypsum	18
Sulphur	98–100
Potassium sulphate	17
Manganese	% Mn
Manganese sulphate ($4H_2O$)	24
Manganese chelates	8–12
Boron	% B
Borax ($10H_2O$)	11
'Solubor'	20
Copper	% Cu
Copper sulphate ($5H_2O$)	25
Copper oxychloride	58
Copper chelates	8–12
Iron	% Fe
Ferrous sulphate ($7H_2O$)	20
Iron chelates	5–12
Molybdenum	% Mo
Ammonium molybdate	53
Sodium molybdate ($2H_2O$)	38
Zinc	% Zn
Zinc sulphate (H_2O)	35
Zinc chelates	10–12
Cobalt	% Co
Cobalt sulphate ($7H_2O$)	20

fertilizers to give 4–8 per cent of magnesium in the final product. Such fertilizers, used on a regular basis, would help to balance the magnesium lost in crops or by leaching from calcareous soils or from acidic soils on which no magnesian limestone is used. Calcined magnesite may also be used as a 'straight' long-term fertilizer. An application of 200–300 kg/ha should help to increase the magnesium content of crops for a period of 3–5 years.

Kieserite ($MgSO_4.H_2O$) 28 per cent MgO and *Epsom salt* ($MgSO_4.7H_2O$) 16 per cent MgO, are water-soluble magnesium fertilizers.

They are more expensive per unit of magnesium than magnesian limestone or calcined magnesite but give more rapid responses where symptoms of magnesium deficiency are found in crops. Kieserite, although more slowly soluble in water, is much cheaper than Epsom salt. An application to the soil of 200–300 kg/ha of kieserite should supply some magnesium to crops for 2–3 years.

Sulphate of potash magnesium, 23 per cent K_2O, 10 per cent MgO and *Kainit*, 10–25 per cent K_2O, 6–20 per cent MgO, supply some magnesium but are used primarily as potassium fertilizers.

Ammonium nitrate with calcium/magnesium carbonate. Ammonium nitrate traditionally mixed with calcium carbonate (Nitro-chalk) may equally well be mixed with dolomite ($CaCO_3.MgCO_3$). A 26 per cent N final product would contain 5 per cent Ca and 3 per cent Mg in the slowly available carbonate form.

Sulphur fertilizers

In the past the fortuitous sulphur content of the then most commonly used nitrogen and phosphorus fertilizers – ammonium sulphate, 24 per cent S, and superphosphate, 11–13 per cent S – along with sulphur reaching the soil from atmospheric pollution, ensured that crops in most areas of the British Isles and other industrial countries received sufficient available sulphur for crops. The clean-air legislation and the switch to ammonium nitrate, urea and ammonium phosphates as the main nitrogen and phosphate fertilizers have radically changed this and it is now necessary to consider using sulphur fertilizers as a routine in the less polluted areas.

Gypsum ($CaSO_4.2H_2O$) 23 per cent Ca, 18 per cent S, is the cheapest form of sulphur fertilizer obtainable in the British Isles. It may be applied to the soil after grinding in the same way as limestone. It is extracted from mineral deposits and is also a major by-product in the manufacture of ammonium phosphates and triple superphosphate. Surprisingly the by-product form is little used as a fertilizer at present. As the demand for gypsum increases, easy-to-spread granular forms will undoubtedly be made.

Potassium sulphate, 50 per cent K_2O, 17 per cent S, is water soluble and if included as the standard potassium source in compound fertilizers would help a good deal to alleviate sulphur deficiency problems.

Commercial sulphur, almost 100 per cent S, has been used directly as a sulphur fertilizer in parts of the world where there is severe sulphur deficiency. A threatened world shortage of this material some thirty years ago seems to have been overcome by the recent discoveries of deposits. It has the obvious advantage of having the highest possible sulphur concentration but has varied very much in its effect on crops. In common with all other nutrient elements, elemental sulphur is not directly available to the plant. It must first be oxidized to sulphur dioxide (SO_2) or sulphurous acid (H_2SO_3) and further oxidized to sulphuric acid or sulphates. Although this can occur by simple chemical processes it is very much dependent, in soils, on the specific action of certain sulphur-oxidizing bacteria. These organisms prefer acidic conditions and thrive in sulphur-rich soils. They may be inactive or even absent in alkaline, sulphur-deficient soils and this is probably the main reason for the variable crop responses to commercial sulphur.

Commercial sulphur is a long-established general purpose fungicide. If applied for this purpose in the British climate it is likely to be washed off the leaves quickly. Used at the recommended rate of about 10 kg/ha this would supply some available sulphur through the soil.

Sulphur dioxide (SO_2) 50 per cent S, must be regarded as a fertilizer, although much of it is cost-free! It is the main sulphur pollutant of the atmosphere and is readily converted in the soil to available sulphates. Without it, many of our crops would be suffering from sulphur deficiency. It would, therefore, be reasonable to use the gas, transported under pressure in cylinders, in the same way as ammonia is used as a nitrogen source. The problems involved would be similar to those encountered with liquefied anhydrous ammonia. There would be a risk in poorly structured, stony or heavy clay soils of considerable losses of sulphur dioxide to the atmosphere. It might, therefore, be preferable to use sulphur dioxide dissolved under pressure in water.

Sodium fertilizers

Sodium is not a major element for most crops but it can be used to replace part of the potassium requirement of crops. The general function of potassium in regulating osmotic pressure within the plant can be performed by sodium but some potassium is always needed for specific functions in enzyme actions and carbohydrate transfer.

In practice, sodium fertilizers, although cheap, are little used except for root crops such as sugar beet, mangolds, turnips, red beet and carrots. *Sodium chloride* (NaCl) is sold as Agricultural Salt, 37 per cent sodium. Its prime use is on sugar beet, where it can replace as much as 50 per cent of the potassium requirement.

There is a good deal of evidence that sodium chloride could be used to replace some of the potassium chloride used for other main crop species, including cereals and grasses, but *not* for potatoes because of their sensitivity to chloride.

Sodium chloride is very cheap and reasons for its limited use include problems of incorporation in compound fertilizers, fears of adverse effects of sodium chloride on soil structure (largely unfounded at the rates required) and the long-established tradition of using potassium chloride in compound fertilizers – ably perpetuated by the sellers of potassium salts on the world market.

Despite this, there could well be a significant role for sodium chloride in compound fertilizers as a cheap replacement for a quarter to half of the potassium, depending on the crop.

Trace element fertilizers

Very few traditional fertilizers, whether simple or compound. contain sufficient trace elements to have any appreciable effect on crops. An exception is basic slag which contains variable amounts of manganese and iron which become slowly available to the plant.

Compound fertilizers have been made containing one or more trace elements but they are not extensively used: routine applications of trace elements are still rare. Trace element fertilizers may be used to correct deficiencies indicated by specific visual symptoms, mainly on the leaves of plants. Some are applied to the soil but, in contrast to major element fertilizers, they are commonly sprayed on to the leaves of growing crops and partly absorbed by them.

Commonly used trace element fertilizers are listed in Table 15.8 along with methods and recommended rates of application.

Foliar sprays   Plants can absorb water and nutrients directly through their leaf pores. Because of the intense competition that the plant must face to obtain nutrients from the soil (see Fig. 4.2, p. 30) it is tempting to short-cut the soil by applying nutrients to the leaves. This has been very successful in the treatment of trace element deficiencies where the quantities of nutrients required are small and one spray treatment is commonly sufficient.

It is much more difficult to use major nutrients in this way.

● The quantities required are such that repeated spray treatments are necessary.
● In the early, critical stages of growth the leaf area of the plant is very small and a large proportion of any spray treatment falls on bare soil.
● Almost all substances used as foliar nutrients are salts, similar to those applied to the soil. These salt solutions drying out slowly on leaves will scorch. Urea, not being a salt, will cause less scorch. Unfortunately, rapid drying on the leaves, which reduces scorch risk, also reduces the efficiency of absorption.
● To achieve maximum absorption through the leaf pores, very fine droplets are needed and this encourages scorch.
● In climates like that of the British Isles opportunities for repeated spray applications without serious risk of the spray being washed off the leaves by rain are limited.

For these reasons foliar application of major nutrients is very restricted in the British Isles.

Unorthodox 'fertilizers'   In addition to fertilizers of proven value the farmer is faced with a range of materials the value of which he does not know. Some are presented with alluring literature making extravagant claims about the product. One such product claimed to increase yield and sugar content of sugar beet, to give a larger proportion of seed-sized tubers in the potato crop and to eradicate every form of insect pest by disrupting their epithelial tissue!

It was a brown fizzy liquid which turned out to be soda water containing dissolved caramelized sugar! This represents the lunatic fringe of the fertilizer world and it sold some fifteen years ago at £5 per litre.

Products range from purely charlatan to respectable but very expensive. Misleading trade names are common among them. Many survive for only two or three seasons in the first of which there is a sales build-up, in the second the big sell and in the third some residual sales, as the true value of the product becomes clear. None of them should be used without prior consultation with a reputable advisory organization.

Chapter 13 Fertilizers and crop yield

Adverse effects on growth and yield

Although the beneficial effects of fertilizers on crop yields are beyond question several adverse effects on plant growth can occur through misuse. These adverse effects are usually associated with water-soluble fertilizers and occur mainly in the early stages of growth. In the worst cases crop yields are reduced.

Most water-soluble fertilizers are, in chemical terms, salts. If they are broadcast and thoroughly incorporated into the soil before the crop is sown they should cause no injury even to tender germinating seedlings. If misapplied – for example by using too much fertilizer or by placing it too close to the seed – serious injury to plant roots may occur, killing seedlings and damaging more advanced plants.

Exchange acidity

When a soil which has not been limed for some time begins to become acidic, hydrogen ions attach themselves to the cation-exchange complex. Similar conditions occur in acidic soils, recently reclaimed for agriculture, which have been limed but in which the cation-exchange complex is not yet dominated by calcium. In these circumstances hydrogen ions (H^+) can be displaced by other cations such as the NH_4^+ derived from newly applied fertilizer and released into the soil solution. This phenomenon sharply reduces the pH and is known as exchange acidity. The released hydrogen ions are eventually leached but in dry spring weather they persist in the soil. As a result toxic amounts of manganese, aluminium and possibly other elements such as copper and nickel are dissolved in the soil solution and taken up by the plant. Affected plants have scorched brown roots. The leaves of cereals may take on a striped appearance of yellow and green, with death of the tips. Seedlings may be killed.

Exchange acidity should not be a problem in calcareous soils or soils which have been well limed for 30–40 years. In these soils the proportion of hydrogen ions on the cation-exchange complex is very small.

Acidity caused by nitrogenous fertilizers

The acidification of soils by ammonium fertilizers should not normally injure crops so long as liming policy is properly adjusted. Where high levels of fertilizer nitrogen are used and liming is neglected injury to plants may occur. In grassland or on soils where direct drilling or minimal cultivations are practised the top 5–10 cm of the soil can become very acidic while lower layers of soil contain adequate lime. In such cases plant roots may be damaged.

Osmotic effects of fertilizers

A high concentration of fertilizer salts in the soil solution can cause severe damage to plants, especially seedlings. This is critical in arid parts of the world where zones of high salt concentration in soils are common. It can also be devastating in temperate humid climates.

Damage is most likely to occur if water-soluble fertilizer is placed in contact with, or close to, seeds in a moist but rapidly drying soil, i.e. sufficient water is available to dissolve the salts but not to disperse them through the soil. Such conditions occur frequently in the British Isles shortly after spring crops are sown.

The main contributors to high salt concentration in the soil solution are chlorides and nitrates. Replacing sulphate by nitrate to produce more concentrated fertilizers has, therefore, increased the risks.

If a root enters a zone of high salt concentration the process of osmosis, by which the plant normally takes up salts dissolved in water, is interrupted. The root becomes dehydrated and scorched. Its efficiency is reduced and, in the extreme, it will die. Results of this are the death of some seedlings and serious setbacks to the growth of surviving plants.

Cases have been recorded in field experiments of fertilizers placed in contact with seed potatoes causing damage sufficient to delay the emergence of shoots by 10–15 days compared with broadcast fertilizer. Affected plants have seriously scorched roots and, although they produce new roots as the salt concentration declines, may take 10–15 weeks to catch up, in terms of total dry matter production, with plants from plots which have received no fertilizer at all.

In a rapidly drying soil placement of fertilizer as far away as 5–10 cm from the seed can cause damage. This is not surprising if one considers that the *effective* salt concentration in the placement zone can be, for row crops, some 20–30 times as great as that of the same amount of fertilizer if it were evenly distributed through the top 25 cm of soil. With a standard application of potato fertilizer the placement zone concentration could be equivalent to a dressing of 30–40 t/ha. In any terms this is a very large dressing and any unwary root, earthworm or other small creature straying into such a zone within a few days of application would be at risk.

The effects of fertilizer salt damage are readily seen in field experiments where comparisons may be made between various fertilizer treatments. In commercial crops, unless there is a complete kill, or 'gapping' in the rows, the damage is insidious because there are not necessarily any leaf or stem symptoms. The crop may look healthy although it has been very badly set back.

Striking examples of fertilizer salt damage occurred some years ago when a type of placement drill, now withdrawn, was used which unfortunately did not place the fertilizer uniformly along the drill. The result was regular gaps in the emerged seedlings corresponding with areas of high salt concentration as determined by conductivity measurements. The gaps remained throughout the season, seedlings having been killed by the concentration of fertilizer.

The dangers of placing fertilizer are much greater than they were thirty years ago because of increase in water solubility and increased rates of application. The only sure way to avoid damage is to broadcast and thoroughly incorporate fertilizer into the seedbed some days before sowing the crop.

Ammonia release

Ammonia gas may be released from ammonium fertilizers or urea, especially if they are applied as top dressings to calcareous soils. Available nitrogen is thus lost to the atmosphere. The escaping ammonia does little damage to the crops in the field but if the release occurs in a still atmosphere, such as that in a closed greenhouse, the ammonia collects immediately above the soil and can seriously scorch the lower leaves of crops. I have seen this on a crop of chrysanthemums – all leaves up to about 15 cm above the soil were killed but the plants survived.

Considerable leaf scorch can occur above the slits when liquefied anhydrous ammonia is injected into difficult soils with standing crops – especially grass. Roots in the injection zone are also killed and grass crops can suffer a serious set back. This problem is more fully discussed on page 128.

Beneficial effects on growth and yield

Provided that early season injury to seedlings or young plants by misused fertilizers is avoided, beneficial effects become apparent very quickly. Countless observations in field experiments have demonstrated this. Yellowing of the leaves of crops on plots which have not been given nitrogenous fertilizers contrasts with the rich green more strongly growing plants on fertilized plots. Most striking of all early season effects is that of water-soluble phosphate fertilizers applied to soils seriously deficient in phosphorus. Treated plots can produce ten or even twenty times as much plant dry matter in the first few weeks of growth as plots to which no fertilizer phosphate has been added. The effects of potassium fertilizers are less obvious early in the season but develop as the crop grows on.

Direct visual evidence in commercial crops is to be seen wherever a small strip has been missed when fertilizer is being applied. The usual appearance indicates nitrogen deficiency with starved, pale green or yellow plants.

Establishment of leaf cover

The early establishment of leaf cover is strongly affected by fertilizer treatments. Increasing the rates of nitrogen and, to a lesser extent, phosphorus and potassium to an optimum level will assure early and efficient leaf cover. This will lead to effective photosynthesis and to maximum transpiration of water at a period of the year when there is usually adequate soil water supply. The longer the resultant efficient leaf canopy is maintained the greater will be the *total* dry matter production of the crop.

The rate of nitrogen required to give optimum leaf cover varies a good deal from crop to crop. If the optimum rate is exceeded the plant will produce an excess of leaves and will be less efficient than a plant grown at the optimum. The reason for this is self-shading. The lower leaves of a plant do not receive sufficient light; they turn pale green or yellow and do not photosynthesize efficiently but continue to respire so that there is a net loss of assimilates to the plant, i.e. less sugars are being manufactured for conversion to starch or proteins. The severity of this effect will vary from crop to crop and within a crop species, according to the density of plant population. It will be more important, for example, in potatoes, which establish a layered canopy of leaves which persists through the season, than in regularly cut grass which has less time between cuts to establish a seriously self-shading leaf system.

A good measure of the likelihood of serious self-shading in a crop is obtained by measuring the leaf-area index which is the total area of leaves divided by the area of ground covered by the plant. There are simple sampling techniques by which this can be measured accurately.

Figure 13.1   Yield and leaf-area index of potatoes as affected by the rate of nitrogenous fertilizer. Unpublished data of P. C. Harper, P. Crooks, R. B. Speirs and K. Simpson.

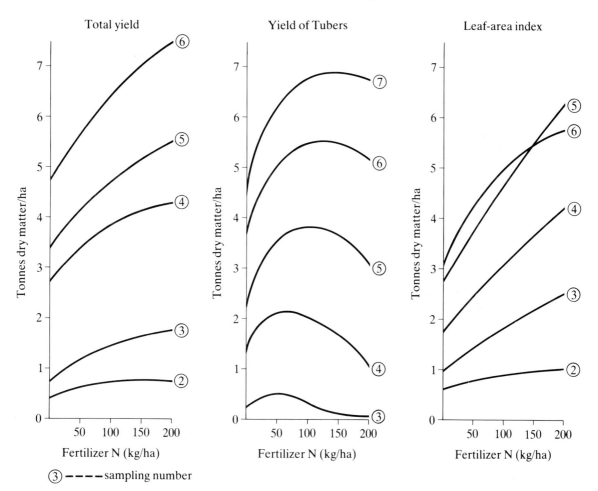

Figure 13.1 shows the effect of increasing the rate of nitrogen fertilizer on the growth and yield of potatoes sampled at intervals

through the season. The three parts of the figure show the total yields, yields of tubers and leaf-area indices at various stages of growth. Samplings are indicated by the ringed numbers. Sample 1 was taken before emergence and samples 2–6 followed at intervals of 3 weeks. Sample 7, tubers only, was at final harvest. The total yield of dry matter (leaves, stems, roots, stolons and tubers) was increased at all stages of growth by increasing rates of N fertilizer up to the greatest amounts used. The yield of tubers, the economic part of the crop, was on the other hand depressed at the higher nitrogen rates as compared with moderate rates. This was very marked in the stages of rapid tuber development 12–18 weeks after planting. The highest yield of tubers at the fourth sampling (14 weeks after planting) was obtained with only 60 kg N/ha, at the fifth sampling (16 weeks) with 105 kg N/ha, at the sixth sampling with 120 kg N/ha and at harvest with 140 kg N/ha. The leaf-area index corresponding to the highest tuber yields at various stages of growth was between 3 and 4.5. 'Excess' nitrogen application resulted in almost double these values and gave not only a reduction in final tuber yield but also delayed tuber initiation and development.

Such effects can be critically important in cooler areas with relatively short growing seasons where all too often the tendency is to increase nitrogen rates in an effort to produce greater yields. They are also important when, for management, climatic or economic reasons, the life of the crop is cut short before maturity. A crop of 'new' potatoes, for example, would be harvested some 12–15 weeks after planting. In the example taken in Fig. 13.1 the use of more than 50–100 kg N/ha would seriously reduce the yield of tubers in such circumstances. In areas such as northern Ireland and eastern Scotland potato crops are grown for seed production. In order to prevent too large a proportion of tubers growing to a size not acceptable as seed, usually about 6 cm diameter, haulms are destroyed between sixteen and twenty weeks after planting. Even at this stage applications of 150–200 kg N/ha would reduce yield as compared with the optimum level. It is therefore important to avoid excess nitrogen applications which help to produce plants with large leaf areas and in which the transfer of assimilates to the tubers is retarded.

Potatoes were taken in Fig. 13.1 as an example. In fact, leaf area as affected by rates of fertilizer application is important in all crops, but especially those which transfer assimilates to storage in roots, tubers or seed during the course of one season. This includes sugar beet, potatoes, turnips, grain crops and grass grown for seed production.

A classical example of the dangers of producing too much foliage by the use of excessive amounts of nitrogenous fertilizers was seen in the sugar beet crop during and for some years after the Second World War in areas of eastern Scotland where the crop is near to its climatic limit. Late sowing was unavoidable (early–mid May) and harvest conditions on many soils were poor after October. In this very restricted growing season, with long day-lengths, the crop tended to produce massive leaf growth with insufficient time for transfer of assimilates to store as sugar in the roots. Optimum nitrogen levels for sugar production were found to be very low (50–100 kg N/ha) but were often grossly exceeded in practice. A common result was the harvesting of only 12–25 tonnes of beet per hectare but greater yields of tops which were difficult to handle and use in the poor harvest conditions. In contrast, the crop grown in East Anglia can come much nearer to maturity before harvest and can, therefore, utilize more fertilizer nitrogen.

Crops which suffer least from excess leaf-area problems are obviously those in which the leaves *are* the crop or a large part of it. Cabbage is a good example of such an arable crop and grass, regularly cut or grazed and effectively producing several leaf crops a year, is the prime example. Such crops will benefit from much higher levels of fertilizer nitrogen than grain, root and tuber crops.

Leaf-area maintenance

The first aim of fertilizer use is to establish as soon as possible an optimum leaf area. The second is to enable that leaf area to persist, effectively photosynthesizing, long enough to take full advantage of environmental conditions such as length of season and autumn soil conditions. Full account must also be taken of crop species and the properties of the selected variety, e.g. early or later varieties of potato.

The problem in leaf-area maintenance is to ensure that sufficient nutrients are available to the crop to sustain it through the grand period of vegetative growth (see Fig. 2.1, p. 14). While available nitrogen and phosphorus have dominant roles in producing a good leaf area, all nutrients are equally important in maintaining it. Deficiency of any of the major or trace elements will lead to early senescence and death of leaves and will reduce crop yield.

If all the fertilizer is applied as a large dressing at or around sowing time there is vigorous early uptake of its nutrients by the plant. This creates the double danger of excessive foliage early in the season and an insufficiency of nutrients later. This is one of the reasons for the increasing use of split dressings for cereal crops.

It seems absurd to apply the whole of the fertilizer requirements of crops such as sugar beet or potatoes before sowing, when they produce as little as 2–5 per cent of their total dry matter in the first 6–8 weeks of growth. This is particularly the case in a period of cold spring weather. During this period the nitrates in the fertilizer will be subject to leaching, some phosphate will be fixed and other fertilizer nutrients will be subject to all the influences that compete with the plant (see Fig. 4.2, p. 30).

In the past there have been technical difficulties in applying fertilizers to some crops during the growing season but the increasing use of tramline systems, the granulation of solid fertilizers and the use of large droplet or dribble-bar application of liquid fertilizers have all helped to reduce damage to crops and to permit uniform mid-season applications of fertilizer.

Effects on final yield

Extensive field experiments and laboratory research have shown that plants have a limit to their response to increasing supply of a nutrient (Fig. 3.2, p. 26). This is reflected in the final crop yield. In field crops each increment of fertilizer gives a progressively smaller increase in yield until a maximum yield is reached. Beyond this point any increase in fertilizer rate will result in either no further increase in yield or a decrease in yield as compared with the maximum. This is illustrated in Fig. 13.3 (p. 155) in which the yield with no fertilizer, Y_0 is that which can be achieved with the amount of nutrient derived from the soil alone and Y_{max} is the maximum yield that can be achieved by adding a particular fertilizer nutrient.

Fertilizer efficiency

The efficiency of a fertilizer in increasing crop yields may be measured in purely biological terms using data such as the gross dry matter yield of the whole plant or the yield of dry matter. This may be called the biological efficiency of the fertilizer. A more useful measure in economic terms uses the increase in value of the crop per unit addition of fertilizer – the economic efficiency of the fertilizer.

Biological efficiency

The biological efficiency of a fertilizer can be expressed in terms of the total dry matter production of the crop per unit of nutrient applied. It depends upon the ability of the plant to absorb the nutrient and to use it, once taken up, for effective dry matter production.

Once in the plant the nutrient will be well used only if the plant is receiving sufficient carbon dioxide, oxygen, sunlight, water and other nutrients to allow it to function efficiently. Any restriction of these can lead to accumulation of nutrients in plant tissue without growth response. A typical example of this is the accumulation of nitrate nitrogen, performing no immediate useful function, in the leaves of a plant suffering from sudden drought or a sudden cold

Figure 13.2   The efficiency of nitrogenous fertilizer in terms of plant dry matter production.

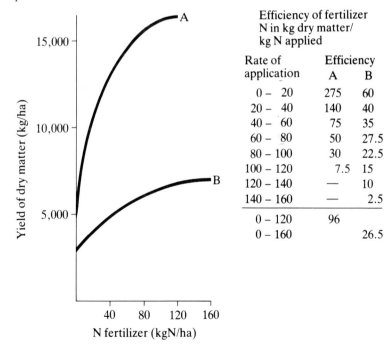

Efficiency of fertilizer N in kg dry matter/ kg N applied		
Rate of application	Efficiency A	B
0 – 20	275	60
20 – 40	140	40
40 – 60	75	35
60 – 80	50	27.5
80 – 100	30	22.5
100 – 120	7.5	15
120 – 140	—	10
140 – 160	—	2.5
0 – 120	96	
0 – 160		26.5

spell. Sometimes the inefficient use of the nutrient by the plant for a short period does not seriously affect final yield although maturity may be delayed. In other cases yield is reduced.

A good measure of the biological efficiency of a fertilizer for a particular crop is the amount of extra dry matter produced for each unit addition of fertilizer nutrient. In these terms small amounts of a fertilizer nutrient applied to a soil deficient in that nutrient are usually very efficient. As the amount of fertilizer is increased each unit applied becomes less efficient until, at the point of maximum yield, the efficiency of further additions of fertilizer is zero or even negative. An example is given showing the effects of increasing rates of nitrogenous fertilizer on the dry matter yield of an unspecified crop (Fig. 13.2). Graph A represents what might happen in a good growing season with no major restrictions to crop growth except an inherent deficiency of nitrogen. In this case the efficiency, as measured in kg dry matter produced per kg of fertilizer nitrogen applied, decreases from 275 at low rates of application to zero at 120 kg N/hectare. The average efficiency for maximum yield is 96 kg dry matter per kg of fertilizer nitrogen.

Graph B represents what might happen if the crop were adversely affected in some way not directly concerned with the nutrient in question. For example the maximum yield attainable might be restricted by temporary waterlogging associated with abnormally high rainfall. In this case efficiency at all rates of application has been reduced, the final yield is much smaller and the average efficiency for maximum yield is only 26.5 kg dry matter/kg fertilizer nitrogen compared with 96 kg for the data in Graph A. Note also that the amount of fertilizer nitrogen required to give maximum yield has been increased while producing a smaller reward.

Economic efficiency and optimum fertilizer rates

In some crops such as grass for conservation there is a close relationship between the total biological efficiency of the fertilizer and its economic efficiency. Effectively the whole crop is taken away and ensiled or otherwise conserved. In most crops the economic part of the dry matter yield is only a fraction of the total. A very good example of this is oil-seed rape in which the economic yield of some 3–4 tonnes per hectare of seed is only a very small part of the total dry matter production including the stems, leaves and roots, which are of little value. Similarly potato haulms are of no economic value, sugar beet tops are usable as animal food only with difficulty and in cereal crops the grain is worth very much more than the straw. Economic efficiency calculations make allowances for these facts by considering the extra value of the crop produced by increasing amounts of a fertilizer nutrient.

Figure 13.3 and Table 13.1 illustrate the calculation of extra profit from fertilizer. As the maximum yield is approached a point will be reached beyond which it is uneconomic to use any more fertilizer nutrient for the crop. The cost of buying, storing and applying any extra fertilizer will then exceed the value of the extra yield produced. This is known as the optimum fertilizer rate (F_{opt} in Fig. 13.3) and the yield at that point is the optimum yield (Y_{opt}) which is always slightly less than the maximum yield (Y_{max}).

On a year-to-year basis the farmer should aim to apply a fertilizer nutrient at a rate as close to the optimum as possible.

Figure 13.3   Nitrogenous fertilizer rates and crop yield.

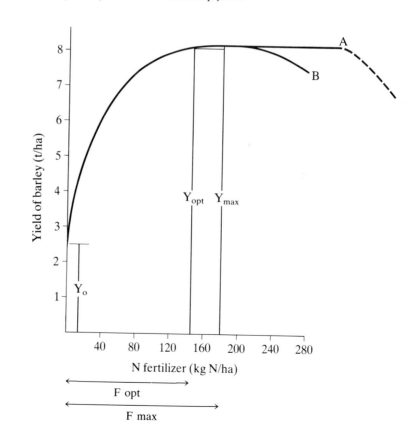

Table 13.1   Calculation of optimum fertilizer rate from the data in Fig. 13.3.

Rate of fertilizer N (kg/ha)	Cost of fertilizer N (£/ha)	a Increase in* cost of fertilizer (£/ha)	Yield of barley (t/ha)	Value of barley (£/ha)	b Increase in* value of barley (£/ha)	c† Extra profit from 20 kg N/ha increment (£/ha)
Nil	Nil	Nil	2.5	305	Nil	Nil
20	8	8	4.8	586	281	273
40	16	8	6.1	744	158	150
60	24	8	6.95	848	104	96
80	32	8	7.45	909	61	53
100	40	8	7.8	952	43	35
120	48	8	8.0	976	24	16
140	56	8	8.1	988	12	4
(150)	(60)	(4)	(8.125)	(991)	(3)	(−1)
160	64	8	8.15	994	6	−2
180	72	8	8.2	1000	6	−2

Conclusion: Optimum fertilizer rate = slightly less than 150kg N/ha.

* Increase per increment of 20kg N/ha   †c = b − a

Penalties of using less than optimum fertilizer rate

As illustrated in Table 13.1, applying less than the optimum fertilizer rate gives smaller yields. The loss of yield is progressively greater for every unit of fertilizer. The financial effects of applying less than the optimum will vary from year to year, soil to soil and crop to crop. In cash crops there will usually be a reduction of profit. In forage crops the ability to use the produce efficiently by adjusting stocking rates and other aspects of management will also play a part. Because of variations in value of crop and cost of fertilizer, Tables 13.2 and 13.3 simply illustrate the effects of using less or more than the optimum level of fertilizer for a cash crop at prescribed values. The yield curves in Fig. 13.3 are used to calculate the rate in Tables 13.2 and 13.3:

● Maximum yield
    8.2 t/ha barley grain.
● Price per tonne
    £122.
● Value of maximum crop
    £1 000/ha (8.2 × 122).

● N fertilizer rate at maximum point
    180 kg/ha.
● Unit cost of N in fertilizer
    £0.4/kg.

Table 13.1 shows the calculations of extra profit from increments of 20 kg/ha of fertilizer nitrogen. The optimum rate of fertilizer is that at which the extra profit from an increment of fertilizer becomes zero. This is just less than 150 kg N/ha.

Table 13.2   Effects on profit (£/ha) of using rates of fertilizer less than F_{max} *

Value of crop £/ha	Cost of F_{max} £/ha	Rate of fertilizer as per cent of F_{max}				
		60	70	80	90	100
	50	−23	−10	−3	−1	**0**
1000	100	−3	+5	+7	+4	0
	150	+17	**+20**	+17	+9	0
	50	−66	−35	−14	−7	**0**
2000	100	−46	−20	−6	−2	**0**
	150	−26	−5	**+4**	+3	0
	50	−109	−60	−29	−13	**0**
3000	100	−89	−45	−19	−8	**0**
	150	−69	−30	−9	−3	**0**

The data in this table are calculated from a yield curve similar to that in Fig 13.3A. The bold figures in each section represent maximum profit from fertilizer, i.e. optimum rate.

*F_{max} represents the smallest amount of fertilizer required to give maximum yield as shown in Fig 13.3.

Table 13.2 is derived from the yield curve in Fig. 13.3 and shows how the profits vary when 10, 20, 30 and 40 per cent less fertilizer is applied than that required at the point of maximum yield. Values of £1 000, £2 000 and £3 000 per hectare are taken for the crop value at maximum yield and the assumed cost of fertilizer to give maximum yield is varied from £50 to £150/ha.

In this table the bold figures indicate maximum profit and this coincides with the optimum rate of fertilizer.

Several important conclusions may be drawn from this table:

● If fertilizer costs are high in proportion to the value of the maximum crop the optimum rate of fertilizer is much lower than that required to give maximum yield.

● If fertilizer costs are small in proportion to the value of the maximum crop, the optimum rate of fertilizer and that required for maximum yield are virtually the same.

● Penalties for using less than optimum fertilizer rates are much smaller for low-value crops than for high-value crops.

● The greatest penalties for using less than optimum fertilizer rates occur when fertilizer costs required for maximum yield are lowest and the value of the maximum crop is highest.

Penalties of using more than the optimum fertilizer rate

Because of the penalties of applying less than the optimum rate (Table 13.2) there is a great temptation to use more than the estimated optimum. This is based on the assumption that increasing the amount of fertilizer above the level required for maximum yield will give no further yield increase but will still give the maximum yield as represented by Model A in Fig. 13.3. If this is true the only penalty for exceeding the optimum rate is the cost of the extra fertilizer used. In other words, the farmer would be insuring himself by using 10, 20 or 30 per cent more than the assessed optimum rate. Fig. 13.3 illustrates the dangers of this assumption.

The Plateau and Parabola Models

For the success of insurance fertilizing the yield curve, after reaching the maximum, must follow graph A. This will be called the Plateau Model. It was assumed for many years that this model applied to all crops in all circumstances – that it was possible to increase the rates of fertilizer nutrients well above those required to give maximum yield without any further increase or decrease in yield. Eventually a point would be reached at which injury to the crop would occur and yields would decline (dotted line, Graph A, Fig. 13.3). Unfortunately, in many cases the yield reaches a maximum and then declines with each further increment of fertilizer nutrient (Graph B, Fig. 13.3). This will be called the Parabola Model. In such cases exceeding the optimum rate will lead to much greater reductions in profit than in the Plateau Model. The penalties of exceeding the optimum rate of fertilizer are illustrated in Table 13.3. During the past thirty years it has become increasingly apparent that the Parabola Model applies to many crops in many circumstances.

Table 13.3 Effects on profit (£/ha) of using rates of fertilizer in excess of F_{max}*.

		Parabola Model				Plateau Model			
Per cent of F_{max}		110	120	130	140	110	120	130	140
Value of crop £1 000/ha									
Cost of	50	−11	−23	−40	−63	−5	−10	−15	−20
fertilizer	100	−16	−33	−55	−83	−10	−20	−30	−40
(£/ha)	150	−21	−43	−70	−103	−15	−30	−45	−60
Value of crop £2 000/ha									
Cost of	50	−17	−36	−65	−106	−5	−10	−15	−20
fertilizer	100	−22	−46	−80	−126	−10	−20	−30	−40
(£/ha)	150	−27	−56	−95	−146	−15	−30	−45	−60
Value of crop £3 000									
Cost of	50	−23	−49	−90	−149	−5	−10	−15	−20
fertilizer	100	−28	−59	−105	−169	−10	−20	−30	−40
(£/ha)	150	−33	−69	−120	−189	−15	−30	−45	−60

The data in this table are calculated from yield curves similar to those in Fig. 13.3.

*F_{max} represents the smallest amount of fertilizer required to give maximum yield as shown in Fig. 13.3.

The penalties incurred for exceeding the optimum assuming the Plateau Model may be acceptable. The Parabola Model, however, shows large penalties as soon as the maximum yield is reached due to decreased yield caused by increasing fertilizer rates. Insurance fertilizing is very inefficient indeed if this model is applicable, especially for high-value crops. The only gain from it, that might be set against the reductions in yield, is some small residual value of the excess fertilizer for subsequent crops. This could never balance out the reduced profit on the current crop.

The important question when deciding whether or not to adopt an insurance fertilizer policy – that is to apply more than the estimated optimum – is 'Which model in Fig. 13.3 applies to this crop, at this site, in this season?' If we had an absolutely firm answer to this question and, as important, if we were certain of optimum rates for all circumstances, the decision would be much easier to make.

There is not yet enough knowledge to say with absolute

certainty which model will apply but good general guidance can be given, especially where all-important decisions on rates of nitrogenous fertilizers are concerned.

Nitrogenous fertilizers   *The Plateau Model* would usually apply where:
● Nitrogenous fertilizers are used for pure grass swards with several cuts per season for conservation or regularly grazed.
● Nitrogenous fertilizers are used for leafy and stemmy brassica crops such as cabbage or kale.
● Nitrogenous fertilizers are used for potatoes, sugar beet, oil-seed rape or cereals *but only* if grown under near ideal conditions (i.e. the growing season allows crops to mature normally, there are no restrictions of excessive dryness, wetness or low temperatures and no pest or disease problems).

The Parabola Model would apply in most circumstances and especially where nitrogenous fertilizers are used for sugar beet, mangolds, potatoes, oil-seed rape and cereals in non-ideal conditions and especially:
● If the length of the growing season is restricted by a late spring, late planting or early harvest requirements (e.g. 'new' or seed potatoes, dates for delivery of sugar beet to factory).
● If there are periods of restricted growth due to drought, low temperatures, temporary waterlogging, pest or disease attacks (i.e. any factors which will prevent the effective transfer of assimilates to the economic parts of the crop – potato tubers, sugar beet roots, cereal grain, oil-seed rape seeds).
● If water-soluble fertilizers are placed too close to seeds, followed by drying conditions, causing injury to seedlings at high rates of application.
● If the soil pH is reduced to a critical level for the crop by large applications of fertilizer.

Phosphorus and potassium fertilizers   *The Plateau Model* will normally apply:
● If water-soluble P and K fertilizers are broadcast and thoroughly incorporated in the soil before sowing the crop.
● If water-insoluble phosphorus fertilizers are used.

The Parabola Model will apply:
● If water-soluble fertilizers are placed too close to seeds, followed by drying conditions, causing injury to seedlings at high rates of application.
● In soils marginally deficient in magnesium, where large applications of potassium fertilizers will induce magnesium deficiency in the crop.

Thus, applying more than the optimum rate of P and K fertilizers may be acceptable provided that they are carefully mixed with the soil and salt effects are avoided. For most arable crops exceeding the optimum rate of nitrogen fertilizer will reduce yields and profit margins.

Lime   The effects of lime must be considered on a longer-term basis than those of other fertilizers. There is no doubt that, in many soils, exceeding the optimum level of lime will lead to trace element deficiencies which, if not corrected, will reduce yields. The Parabola Model will therefore apply especially in soils prone to trace element problems – for example, sandy soils with low organic matter contents.

Other types of yield response

Continuous slight increase   Most crops in most conditions will conform to one of the two models shown in Fig. 13.3. Less commonly, the yield increases at relatively low fertilizer rates, as in Fig. 13.3, but then continues to increase slightly even at very high rates of application. There is, for example, some evidence from field experiments in England that potato yields continue to increase slightly over a considerable range of fertilizer phosphorus rates. This has been sufficient for ADAS to recommend very high levels of application, up to 350 kg P_2O_5/ha, for this crop. Similar experiments in Scotland have shown no such trend. As a result 'standard' recommended rates of fertilizer phosphorus for potatoes in south-east Scotland are substantially lower than in England, the highest recommended rate being about 200 kg P_2O_5/ha. This difference probably results from the greater release of organic phosphorus from the Scottish soils which contain, on average, more total organic matter.

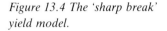

Figure 13.4 The 'sharp break' yield model.

Basic yield patterns

The 'sharp break' model   Another type of yield response, once very common in cereal crops, shows a sharp break in yield as the rate of nitrogenous fertilizer is increased (Fig. 13.4). In cereals this is associated with lodging. The stems of the plant break or bend, the flattened crop suffers from delayed maturity and harvesting is very difficult, often with large losses of grain. In such cases the break in the yield response is caused by the catastrophic effect of excess nitrogenous fertilizer on harvestable yield of grain. It is little comfort to the farmer that if the yield of the lost grain were added to the harvested yield he would have had a bumper crop.

Recent introductions of stiff-strawed varieties and the use of growth retardants such as chlorocholine chloride (CCC) have reduced the incidence of lodging and consequently increased the optimum rates of nitrogenous fertilizer.

It must be stressed that the 'sharp break' Model will apply only in cases where some catastrophe has befallen a crop which has been pushed to breaking-point. Other good examples are severe yield reductions caused by antagonistic effects as shown in Fig. 4.3 (p. 37) for the potassium/magnesium antagonism and in Fig. 5.2 (p. 55) for induced manganese deficiency by overliming. In both cases the antagonism reaches a critical level at which the plant cannot take up sufficient of the antagonized element to maintain itself and there is a sharp reduction in yield.

Attempts have been made to use the 'sharp break' Model to interpret yield responses where no acute symptoms can be seen in the crop (i.e. lodging in cereals, manganese deficiency symptoms or magnesium deficiency symptoms in the three examples taken).

There is, however, overwhelming evidence that the yields of most crops with increasing fertilizer rates follow smooth curves of the types shown in Fig. 13.3 and may be represented by either the Plateau Model or the Parabola Model.

Figure 13.5 shows four basic patterns of crop yield with increasing rates of a fertilizer nutrient and illustrates the wide variations in fertilizer efficiency that occur in practice. Almost all conceivable effects of fertilizers on yield fit one or other of these patterns:
● Low optimum rate – high reward (Graph 1)
● High optimum rate – high reward (Graph 2)
● Low optimum rate – low reward (Graph 3)
● High optimum rate – low reward (Graph 4)

Figure 13.5   Basic yield patterns with increasing fertilizer rates.

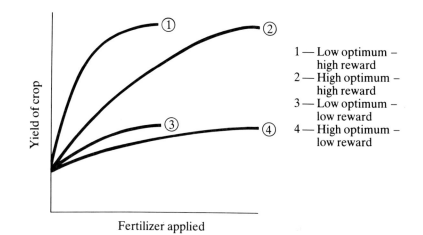

1 — Low optimum –
      high reward
2 — High optimum –
      high reward
3 — Low optimum –
      low reward
4 — High optimum –
      low reward

NB: In Fig. 13.5 it has been deliberately assumed for clear presentation that the yield with no fertilizer is the same for all four patterns. In practice examples could be found showing wide variations in the yield produced without fertilizer. For example the low optimum–low reward pattern (Graph 3) is typical of a soil rich in the nutrient concerned and the yield with zero fertilizer would be much higher than that shown.

The first of these basic patterns, low optimum rate–high reward (Graph 1), is obviously the most desirable. Some important examples of cases in which the four models apply are given below.

Low optimum rate–high reward (maximum fertilizer efficiency)

● Crops grown on soils in excellent physical condition in which roots can tap large volumes of soil. As a result plants can take up fertilizer nutrients efficiently and give high yields for small fertilizer input. This pattern would apply to N, P and K in these conditions but especially to phosphorus which is less mobile in the soil.

● Nitrogen applications for 'new' or seed potato crops.

● Phosphorus applications to low-demand crops such as sugar beet grown on phosphate deficient soils.

● Phosphorus applications to crops such as rape (*Brassica napus*) which utilize fertilizer phosphorus more efficiently than most crops.
● Potassium applications to low-demand crops, such as cereals, grown on potassium deficient soils.

High optimum rate–high reward
● Crops with a high demand for a particular nutrient especially if grown on soils deficient in that nutrient. Examples: nitrogen for kale or pure grass swards, potassium for sugar beet, phosphorus for potatoes.
● Nitrogen applications to soils leached by heavy rainfall in winter and early spring. Example: cereal crops responding to top-dressed nitrogen fertilizers because seedbed applications have been leached to some extent in a wet season.
● Crops grown on soils which have some limiting physical characteristics but not sufficient to give serious yield reductions. In such soils, for example those with blocky structures or cloddy tilths, both phosphorus and nitrogen fertilizers will greatly assist root development but inputs for optimum yield will be high.

Low optimum rate–low reward
● Crops grown on soils in which yields are seriously impaired by acidity. Both the optimum rate and the yield will be low because of root damage which makes it impossible for the crop to absorb and utilize fertilizer nutrients.
● Nitrogen applications for crops attacked by fungi such as botrytis or mildew which are encouraged by high nitrogen inputs (unless adequate corrective measures are used).
● Crops which have received high inputs of nutrients from farmyard manure or slurry. With heavy applications the optimum rate of fertilizer potassium for many crops is reduced to zero.
● Nitrogen applications to leguminous arable crops (peas, beans) or to grass/clover swards rich in clover.
● Nitrogen applications to crops grown on soils in the years immediately after ploughing in clover-rich leys of 2–4 years duration.
● Nitrogen, phosphorus and sulphur applications to crops grown on soils rich in organic matter with moderate–high pH values

(greater than 6.0) in warm, moist seasons when mineralization of these elements is vigorous.

High optimum rate–low reward (minimum fertilizer efficiency)

● Nitrogen applications to crops the roots of which are attacked by nematodes or fungi. These crops can be encouraged by nitrogen fertilizers to produce new roots but yield responses are usually very limited.

● Nitrogen applications to crops affected by waterlogging. Some root production will be encouraged in the top few centimetres of soil but yield responses are limited.

● Crops damaged in early season by salt effects or exchange acidity caused by placing fertilizer too close to the seed in moist soil followed by drying conditions.

The examples given above will not apply universally but are certainly applicable in most cases. There are, of course, many intermediate cases in which moderate optimum rates of fertilizer may give high, moderate or low rewards. Interpretation of yield response patterns, fertilizer requirements and the diagnosis of the reasons for them make a fascinating, useful and profitable study for the practising farmer.

Assessment of optimum fertilizer rates

Optimum fertilizer rates vary widely from crop to crop, season to season and site to site. A barley crop, for example, may need only 50–60 kg P_2O_5/ha in fertilizer form whereas a potato crop grown in the same field may need twice that amount. A sugar beet crop grown on a potassium deficient soil can require as much as 200 kg K_2O/ha whereas a crop grown on a soil with high available potassium may need only 75 kg/ha. A very wet spring may greatly increase the optimum level of fertilizer nitrogen because of intense leaching losses.

The assessment of optimum fertilizer rates is a very complex process. It has involved many years of practical experience handed down from generation to generation of farmers and a century of carefully recorded field experiments in which N, P and K fertilizers have been applied at various rates and in various combinations.

The result has been the publication of tables of fertilizer recommendations by government-sponsored organizations such as the Agricultural Development and Advisory Services in England and Wales (ADAS), the Scottish Agricultural Colleges in Scotland and similar organizations in other countries. Similar tables are produced by many fertilizer-selling organizations.

The tables take account of many of the factors affecting optimum rates and, as a result, there are no set recommendations for a particular crop. As an example, the present ADAS recommendations for winter barley vary between nil and 160 kg N/ha, nil and 80 kg P_2O_5/ha and nil and 80 kg K_2O/ha according to location, soil type, soil analysis, previous cropping, manuring and fertilizer history and the expected yield of the crop.

Factors affecting fertilizer requirements The main factors affecting fertilizer requirements may be divided into four groups:
- Environment – soil, climate, latitude, topography.
- Farming system and management.
- Crop.
- Choice of fertilizers and other substances applied to the crop (see Chs 14–16).

Environmental factors **Soil type** The type of soil has major effects on fertilizer requirements through its physical, biological and chemical properties.

Soil texture has its main effect on the degree of leaching, particularly of nitrate from nitrogenous fertilizers. This is readily lost from light sandy or gravelly soils but more slowly from heavier clay soils because of the lower rate of percolation. Leaching in sandy soils is offset to some extent by the ease with which roots can penetrate and ramify but requirements of fertilizer nitrogen and potassium are generally high.

Good soil structure, encouraged by adequate levels of soil organic matter, can modify leaching losses and increase the ease with which the plant can extract fertilizer nutrients from the soil. Fertilizer requirements on poorly structured soils with pans or large structural units are usually high and crop responses are poor.

Organic matter content affects fertilizer requirements through its influence on cation-exchange capacity and water retention by the soil. Soils containing appreciable amounts of organic matter (2.5–10 per cent) will suffer less from leaching and will hold more fertilizer nutrients in available form than low-organic soils. This will reduce fertilizer requirements.

There is also an annual release of nutrients, especially nitrogen, phosphorus and sulphur, from soil organic matter by mineralization. The actual amounts of nutrients released are dependent on soil temperature, moisture, soil pH and the quantity of organic matter in the soil. Because of this it is not easy to predict the reductions in fertilizer requirements made possible from season to season.

Inherent chemical fertility resulting from rich parent materials, lack of leaching and generous past applications of lime, fertilizers and manures reduces fertilizer requirements. It is reflected in soil analysis results and adjustments to fertilizer rates may be made accordingly.

Climate *Excess rainfall* increases fertilizer requirements, especially of nitrogen and potassium, in freely drained soils by enhancing leaching. It also impairs the efficiency of fertilizers in poorly drained soils because of waterlogging which can cause denitrification, asphyxiation of roots and the release of toxic gases in the soil.

Inadequate rainfall leading to drought or at least to insufficiency of water for maximum plant growth can drastically reduce the uptake of fertilizer nutrients by the plant. Unfortunately, attempts to correct this by using more fertilizer nutrients are likely to result in increased salt concentration which can damage or kill seedlings or young plants.

Soil and atmospheric temperatures have both seasonal and day-to-day effects on the effectiveness of fertilizers and, therefore, on fertilizer requirements. Obviously continuous excessive temperatures will encourage drought and will lead to similar problems to those described above.

The all-important influence of temperature on the length of the growing season affects the ability of the plant to complete the cycle of absorption of nutrients, efficient use of them and effective maturation. In particular a short growing season will generally necessitate reductions in the rates of nitrogenous fertilizers and not, as might be expected, increases.

Latitude and topography Many aspects of soil properties, and especially organic matter type and content, are influenced by latitude and topography and, through them, the amounts of fertilizer that are needed. The cooler conditions induced by increasing altitudes or latitudes reduce potential crop responses to fertilizers.

Topography, especially the slope and aspect of the site, affects the soil temperature, drainage and hence the amount of leaching which will occur. In the northern hemisphere north-facing slopes are appreciably cooler than south-facing slopes. Thus there is less evapo-transpiration of water. A larger proportion of incident rainfall percolates through the soil and leaching is enhanced. This will indirectly affect both the optimum rates of fertilizer nutrients and the efficiency with which they are used by the plant.

Farming systems

Farming systems have a major effect on fertilizer requirements. So much so that, in the east of Scotland, broad categories of farming systems are used as the main groupings in deciding upon fertilizer recommendations.

The groupings are:
- Intensive cash-cropping farms – these farms lie mainly on the eastern seaboard with annual rainfall less than 750 mm.
- Arable farms with approximately one-quarter of the cropping area in rotational grass, including a livestock enterprise – most of these farms lie inland (further west), with higher rainfall levels of 750–900 mm/year.
- Grass/arable farms with at least one half of the cropping area in rotation grass – these farms lie mainly still further inland and westerly and at higher altitudes with annual rainfall of more than 900 mm.

It is accepted that there will be some farms falling into each category that are anomalous. Allowance is made for this by adjustments to the recommendations after discussions between the farmer and his adviser. For the most part the type of farming and the all-important proportions of grassland in the rotation are dictated by the wetness or dryness of the climate.

The recommendations which vary most from category to category are the fertilizer nitrogen rates. These are very strongly affected by the proportions of grassland in the various systems, the consequent variations in livestock numbers and, above all, the organic matter returns to the soil.

The wide variations in recommended rates of fertilizer nitrogen for the three groups are shown in Table 13.4. The range and average organic matter contents of these groups of farms are

Table 13.4   *Effects of farming system (East of Scotland) on recommended rates of fertilizer nitrogen for winter barley and seed potatoes.*

Crop	System A*	System B	System C
Winter Barley	90 – 175	80 – 155	70 – 110
Seed potatoes	140	100 – 130	60 – 80

*A   Intensive cash cropping.
 B   Arable with approximately one-quarter of the cropping area in rotational grass, including a grazing-livestock enterprise.
 C   Grass-arable with at least one half of the cropping area in rotational grass.

Source: Adapted from *Fertilizer Recommendations*, Bulletin No. 28, East of Scotland College of Agriculture.

shown in Table 13.5. They have been found by long experience and experimentation to affect the fertilizer nitrogen requirements of crops.

Farm management

Many aspects of farm management affect fertilizer requirements since many inefficient farming practices lead to less effective fertilizer use. The most obvious cases are neglect of liming and failure to drain the land adequately or to maintain drains. Inadequate pest and disease control renders fertilizers less efficient and may lead to higher optimum rates with smaller rewards in terms of crop yields. Excessively early applications, particularly of nitrogenous fertilizers, large applications before sowing and failure to use split dressings where appropriate can all lead to excessive leaching or fixation losses because of the lapse of time before the crop can take up the nutrients. The formation of cultivation, plough pans or surface-compacted layers in the soil

Table 13.5   *Organic matter content of soils under various farming systems (East of Scotland)*

System	Per cent organic matter	
	Average	Range
Intensive arable	3.8	2.0–6.2
Arable with 25% grass	4.4	1.9–7.4
Arable with 50% grass	5.0	3.6–6.8
Permanent pasture	8.5	7.9–9.5

Source: Data supplied by R.B. Speirs.

will lead to poor use of fertilizer nutrients by the plant and thus to higher optimum rates with poorer rewards in crop yields.

Place of crops in rotation   The place of a crop in the rotation has a strong influence on fertilizer requirements. In traditional rotations cereal crops following well-fertilized and clovery leys of 2–4 years duration have very low requirements for fertilizer nitrogen and phosphorus. The fertilizer nitrogen requirements of subsequent cereal crops increase from year to year. The third cereal crop after ploughing a rich 3-year grass ley will need more than one and a half times the fertilizer nitrogen needed by the first crop after the ley. Similar adjustments must be made for crops other than cereals.

Increasing areas of land are given over to continuous cereal production and all-important crop residues such as stubble and straw are commonly burnt. As a result organic matter levels decline and these soils supply progressively less and less nitrogen for crops from year to year and fertilizer nitrogen rates must be increased. This problem is most severe in the dry areas of south-eastern England where organic matter is readily lost from the soil by oxidation.

Use of farmyard manures and slurries   The optimum rates of N, P and K are all reduced for crops following applications of farmyard manures or slurries. The reductions that can be made vary from crop to crop and according to the amounts and types of manure used and especially the time at which it is applied and incorporated into the soil. Averages and ranges of NPK fertilizer equivalents for farmyard manures and slurries are given in Table 8.1 (p. 89). If the guidelines for making, storing, applying and incorporating manures and slurries set out in Chapter 8 have been closely followed, reductions in fertilizer N, P and K can be made according to the highest figures in the ranges.

Residues from previous manures and fertilizers   Residual effects of phosphorus and potassium from fertilizers or manures tend to build up over the years, especially if excessive amounts have been used. Such a build-up of residues is slow and the use of excessive amounts of water-soluble phosphorus and potassium fertilizers specifically for this purpose is certainly not to be recommended. The actual amounts of residual phosphorus and potassium recovered by crops in years subsequent to the year of application are very small, commonly of the order of 1 per cent of the amount originally applied. Phosphorus and potassium residues are usually reflected in soil analysis results and adjustments to rates of application for the current crop can be made according to scales similar to those presented in Table 15.2 (see p. 189).

Residual nitrogen from previous applications of farmyard manures and fertilizers is more difficult to assess. Only about 10 per cent of that not taken up by the first crop finds its way into the second crop after application. The main residues from fertilizer nitrogen are found not in inorganic forms but in the soil organic matter or in manures made from the crops produced. Allowance can be made for those residues by methods illustrated in Tables 15.5 and 15.6 (see p. 192 and 194).

Choice of crop As shown in Table 3.1 (see p. 22) the requirements of different groups of crops for nitrogen, phosphorus and potassium vary considerably. This is reflected in fertilizer requirements. Because of the varying needs of the crops for particular nutrients, for example the small requirements of phosphorus for sugar beet and potassium for cereal crops, the ratios as well as the amounts of $N : P_2O_5 : K_2O$ are influenced.

To meet this need manufacturers have produced 'standard' fertilizers for particular crops, for example grain fertilizers with a ratio of 3:1:1 or 2:1:1 and potato fertilizers with ratios of 1:1:1.3– 1:1:1.6. There are, however, many cases in which the requirements of crops are not met by these standard fertilizers and it is best to take the implication of suitability for a particular crop as a very general statement and to be prepared to use another fertilizer more suitable to your true requirements.

Expected crop yield Account is usually taken of the expected crop yield when devising fertilizer recommendations. In the past it was sufficient to make recommendations, particularly for phosphorus and potassium fertilizers, based on a single assumed yield. Because of the rapid increases in *potential* yield of some crops, especially cereals, and the consequent wide range of yields from farm to farm, the most recent ADAS fertilizer bulletin recommends different N, P and K levels for a range of expected yields.

Table: 13.6   ADAS recommendations of phosphorus and potassium fertilizer rates for cereal crops.

Recommendations of P_2O_5 (kg/ha)

P index	0	1	2	3	over 3
Expected yield (t/ha):					
5.0	90	40	40(M)*	40(M)	Nil
7.5	110	60	60(M)	60(M)	Nil
10.0	130	80	80(M)	80(M)	Nil
K index	0	1	2	3	over 3
Expected yield (t/ha):					
5.0	80	30	30(M)	30(M)	Nil
7.5	95	45	45(M)	45(M)	Nil
10.0	110	60	60(M)	60(M)	Nil

*M indicates a maintenance dressing to prevent depletion of soil reserves rather than to give a yield response.

Source: From *Lime and Fertilizer Recommendations No.1, Arable Crops 1985/86*, Booklet 2191. ADAS, Ministry of Agriculture, Fisheries and Food.

Table 13.6 gives, as an example, the P and K recommendations at various P or K indices. It is assumed that straw from the previous crop has been ploughed in or burnt.

Chapter 14   Fertilizers and crop quality

The term 'quality' with respect to crops grown for human consumption is an emotive one. It is bandied about freely by the puzzled consumer who suspects that the flavour and food value of foodstuffs are not what they used to be. The manufacturers of processed foods are an obvious target for customer fury but in the case of fresh vegetables, especially potatoes, the farmer and his use of 'artificial fertilizers, becomes the butt of the attack and rebuttal is not easy because of the complexity of 'quality'.

Whether crops are produced for human consumption or for feeding to animals their quality has four main aspects:
- Nutritional value.
- Processing value.
- Storage potential.
- Appearance, flavour, smell and texture.

The four aspects are to some extent interdependent.

Nutritional value

The amounts of extra carbohydrates, fats, oils and proteins that can be produced, per hectare, by the judicious use of fertilizers are quite staggering. The individual human being or animal can, however, only eat so much food per day and the nutritional value of the actual intake then becomes important. Human foods are so many and variable in the wealthy parts of the world and eating to excess is so common that the nutritional content of the diet components is important mainly in respect of preventing obesity.

In countries where undernourishment is a problem both the quantity and quality of the foods produced become important. In such countries the use of fertilizers is seldom excessive and the nutritional quality of foods will usually benefit from any fertilizers used.

It must be stressed that carefully-used fertilizers will generally improve the nutritional quality of the *dry matter* of crops in terms of both protein value and the energy that the animal can derive from the crop. Increasing the amount of any fertilizer nutrient from a deficiency level to optimum level will improve the nutritional quality of the dry matter. It is when excessive fertilizer rates, especially of nitrogen, are used that adverse effects are likely.

Dry matter content

One of the main effects of fertilizers is on the dry matter content of the produce. In general, increasing the rate of either nitrogenous or potassium fertilizer will decrease the dry matter content of the produce whereas increasing the rate of phosphorus fertilizer will increase it. A striking example of the effects of nitrogen and potassium fertilizers is given in Fig. 14.1 for both potato tubers and fresh grass before ensiling.

This dilution effect, although important, is often neglected in considerations of the effects of fertilizers on the nutritive value of

Figure 14.1   Effects of N and K_2O fertilizer rates on the percentage dry matter in fresh grass and potato tubers. Data from lecture given to Food and Agriculture Groups of Society of Chemical Industry, Feb. 1965. K. Simpson, P. Crooks and P. C. Harper.

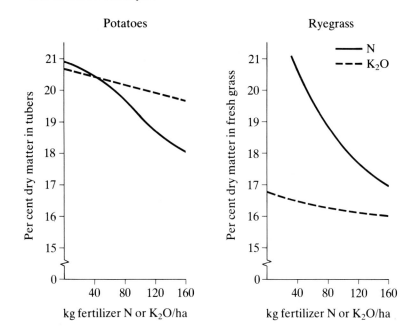

crops. It is not a major problem in cereal grains which are allowed to desiccate mainly in the field and then in the grain drier. In other crops which are consumed fresh and fed directly to the animal there is a simple dilution effect of the extra water in the plant on the nutritional quality of the foodstuff. The animal must eat proportionately more food to get the same food value and the dry matter intake of the animal is adversely affected. Fodder crops such as turnips and especially grazed grass, treated with high levels of nitrogenous and potassium fertilizers, are the best examples.

In terms of human nutrition anyone eating potatoes such as were grown in the experiment shown in Fig. 14.1 would have to eat some 10 per cent more of those grown with 150 kg N/ha than those grown with no fertilizer to get approximately the same nutritional value. They would also have paid for 24 more grams of water in each kilogram of potatoes bought! The potato processing industry is very much aware of this problem which increases their costs in removing the water either by desiccation or frying. Rigid standards are imposed in terms of variety of potato and dry matter content.

Composition of dry matter

Other effects of fertilizers on the nutritional quality of crops concern not only the main groups of substances involved in animal nutrition, carbohydrates, proteins and fats, but also vitamins and minerals required by the animal. It is not possible in this book to deal comprehensively with this subject but some important examples have been selected.

Non-protein nitrogen When a plant is not functioning well, perhaps because of a spell of cold weather, the production of true proteins by the plant is restricted. Nitrogen compounds other than proteins then accumulate in the plant, including amides, amines and nitrates. When the weather improves these substances are usually converted to protein but grass harvested during these periods contains nitrogenous materials that the animal cannot utilize and which may be harmful to it. This reduces the nutritional quality of the grass and is encouraged by excessive fertilizer nitrogen.

In some circumstances nitrate accumulation in the plant tissues can be substantial. If the concentration in the plant exceeds 40–50 mg/kg there is a risk of a consequent build-up of nitrate in the leaves of the plant or in the alimentary tract of animals eating the material. This can cause methaemoglobinaemia – a condition that renders haemoglobin incapable of its essential oxygen- transporting function. There is some risk of this condition occurring in human beings, especially infants, after eating fresh or canned vegetables grown with excessive fertilizer nitrogen and thus containing nitrates or nitrites.

Minerals Fertilizers affect the mineral content of all crops, especially during periods of rapid growth. The mineral elements most affected are those not normally included in fertilizers. In various circumstances magnesium, manganese, sulphur, copper, cobalt and iodine – all essential in animal nutrition – can be affected if amounts in the soil are approaching deficiency levels. The general effect is a reduction of the concentration of these elements in the crop as rates of fertilizer application are increased.

The effects described are not usually important in human nutrition because of the various types and sources of food in the diet but they necessitate careful attention to the mineral nutrition of stock fed mainly from intensively fertilized grass or other crops produced on the farm.

Magnesium. High levels of potassium fertilizer reduce the magnesium content of herbage and may help to produce grass with such a low magnesium content that animals on a diet largely composed of the grass will develop low blood magnesium concentration – hypomagnesaemia – and may die of hypomagnesaemic tetany. This condition is most prevalent in animals eating early season grass and it is wise to avoid the use of potassium fertilizers in early fertilizer applications for spring grass.

Liming with calciferous limestones will also reduce the magnesium content of herbage. Where there is a risk of magnesium deficiency magnesian limestone should be used instead.

Sulphur. The total sulphur content of crops is depressed by using nitrogenous fertilizers such as urea or ammonium nitrate which contain no sulphur. In particular the amounts of sulphur amino-acids, essential to the animal, are reduced in the crop. This could be important in animal production unless the stock are given supplementary sulphur.

Trace elements. Overliming and even moderate liming will reduce the concentration in fodder crops of trace elements essential for the animal – cobalt, copper, iron, zinc and manganese – in some cases to very low levels. It is likely, but not yet proven, that the selenium content of crops is also reduced

Vitamins Definite information about the effects of fertilizers on the vitamin content of crops is sparse.
The indications are that:
- Excessive nitrogen rates reduce vitamin C content.
- High nitrogen rates increase vitamin B and carotene contents.
- High rates of phosphorus increase the concentration of vitamin A, various B vitamins and vitamin C in plant tissues.

Processing value

Many crops are grown specifically for processing for human consumption – sugar beet for sugar, barley for malting, some types of wheat for bread, others for biscuit making, others for pasta, potatoes for a vast range of processed products, vegetables and soft fruits for freezing or canning. In many cases processors are willing to pay a premium for crops grown to meet their specifications. Commonly the requirements are stringent. This is a large part of the food industry and there is no doubt that fertilizer use can change the acceptability of a crop for a processor.

Barley for malting

A vast amount of barley is used for malting each year. The maltster specifies several properties and pays premiums of 20–30 per cent of the non-malting price. Both premiums and specifications vary from season to season. In a season producing generally poor malting barley the specifications are relaxed. The main chemical specifications are low crude protein, commonly less than 10 per cent (1.6% N) in the grain, and high starch content. The two properties are complementary and are strongly affected by fertilizer nitrogen rate.

Malting quality is so important that plant breeders have produced varieties specifically for malting: Golden Promise, Maris Otter and more recently Midas, Triumph and Kym which tend to have low-protein grains. These varieties, while commonly preferred by maltsters, by no means dominate the market. With judicious use of nitrogen fertilizers and good luck, acceptable crops of other varieties can be grown.

The grower is faced with a difficult decision: to apply nitrogen rates well below those required for optimum yield, which will greatly improve the prospects for a good malting sample; or to apply optimum nitrogen rates, obtain higher yields and hope that a kind season and good harvest will give a well-matured high-starch crop which will be acceptable to the maltster. Many take the latter course.

Nitrogen rates which exceed the optimum rate will seldom produce a crop acceptable for malting. It is also inadvisable to top-dress malting barley with nitrogenous fertilizer more than two weeks after the plants have emerged. In cool, wet areas all nitrogenous fertilizer should be applied either shortly before or immediately after the crop is sown.

Wheat

Wheat for baking bread Like maltsters, bakers apply stringent requirements for wheat and pay premiums for suitable crops. The requirements for bread wheat differ considerably from those for wheat used in biscuit or cake making and yet again for pasta. The basic requirements in bread wheat are high gluten and low α-amylase.

Gluten is a complex formed when wheat-grain protein is mixed with water. It is elastic and also retains the carbon dioxide, evolved during baking, as bubbles. This gives the bread a good structure. Increasing the protein content of the grain will also increase the gluten. This requires as high a nitrogen input as the plant can tolerate without lodging, which may be equivalent to the rate required for optimum yield or even greater, especially if a growth retardant such as CCC is also applied. Late top-dressings of the crop with nitrogenous fertilizer and even foliar applications of urea will help to boost grain protein.

The enzyme α-amylase converts starch to sugars. During bread baking this is undesirable because it causes stickiness, poor structure and unevenly shaped loaves. Some wheat varieties are naturally low in α-amylase, so much so that it is necessary to add some to the mix before baking bread. There is no firm evidence that α-amylase activity in the dough is affected by the amount of nitrogenous fertilizer applied to the crop which produced the grain from which the flour is made. Bread wheat was little grown in the British Isles until recently but the cost of imported grain from Canada and elsewhere has stimulated attempts to grow satisfactory bread wheat here.

Some varieties have been produced by the plant breeders

specifically for bread making. Varieties such as Flanders and Avalon have been grown successfully in England and Maris Huntsman is useful for mixing with imported wheats for bread making.

Biscuit wheat In contrast to bread flour, that used in biscuit making must have a low protein content. This is more easily produced in Britain than bread wheat. The varieties Brigand, Longbow and Norman are useful biscuit wheats. As for malting barley, the use of sub-optimal nitrogen fertilizer rates and the application of all the nitrogen immediately before or immediately after sowing will assist in producing biscuit-quality grain.

Pasta wheat To make spaghetti and other pasta a special type of wheat is preferable. It is called Durum, because the grain is very hard and is suitable, when ground and wetted, for the extrusion processes common to all pasta making. Processors require a high protein Durum wheat, at least 12 per cent crude protein, and to produce this late top dressings of nitrogen should be used as for bread wheats. The varieties Capdur and Valdur are satisfactory and may command a premium of £50 per tonne over bread wheats.

Sugar beet

Sugar beet roots are used exclusively for processing to the various grades of sugar (sucrose). Both the pulp remaining after sugar extraction and the leaves left at harvest are useful as animal food although difficult to handle.

The quality of beet for processing depends upon its sugar content, the ease with which it can be extracted and, to a lesser extent, the amount of soil incidentally carted to the factory. The beet factories will pay a premium for high sugar content, using 16 per cent of the fresh beet weight as a standard, and may exert a penalty for low sugar content. They also frown upon soil contamination because the beet must be washed free from soil before processing. The factories also exert control over the dates on which beet must be delivered to them.

Both nitrogen and potassium fertilizers, especially if used at high rates, strongly influence the processing quality of sugar beet. Both tend to reduce the sugar content of the beet and, in case of nitrogen, this is especially so if early delivery to the factory is required before the crop is mature. This effect is less marked in the relatively warm and dry soils of East Anglia than in wetter, cooler areas. High rates of nitrogenous and potassium fertilizers also affect the composition of the expressed juices of the beet from which sugar is extracted. High nitrogen rates increase the amounts of betaine and α–amino compounds present and high potassium rates give a potassium-rich extract. Both factors make more difficult the production of pure taint-free sugar and the presence of betaine reduces sugar yield.

Boron deficiency, induced by overliming, may reduce the sugar content of beet well below the standard 16 per cent level. Boron deficiency also leads to the death of the central leaves in the crown, sometimes followed by bacterial rots, and to a proliferation of small leaves resulting in harvest problems.

Oil-seed rape

High levels of fertilizer nitrogen depress the oil content of the crop. As with malting barley, the grower must reach a compromise between higher yields of lower quality and lower yields of a higher quality crop grown with a smaller fertilizer input. This makes the basis of payment critical in decisions on fertilizer rates. In practice, in the United Kingdom much of the crop is grown under contract to specifications made by the buyer with regard to variety.

Potatoes

Nitrogen and potassium fertilizers reduce the dry matter content of tubers (Fig. 14.1). This affects not only the nutritional quality but also the quality of the tubers for processing. One of the main problems in processes such as potato crisp manufacture or dehydration is the amount of energy needed to remove water from the tubers. Because of this manufacturers of crisps, potato straws and similar products specify high dry-matter tubers. Varieties such as Bintje (before 1945) and more recently Record, which is still widely grown, are used for processing because of their high dry matter content. This can be reduced to less than 20 per cent, however, if high rates of N and K fertilizer are used, and may give crops that are unacceptable to the processor. Phosphate fertilizers up to optimum rates tend to increase tuber dry matter content and thus improve crisping quality.

A low concentration of reducing sugars in the tuber is also required by crisp manufacturers but this is affected much more by storage conditions than by fertilizer regimes.

Requirements for shape and eye depth are met by specific varieties and are little influenced by fertilizer rates. The ideal potato for canning, for example, is rounded, small in size and has shallow eyes. The variety Maris Peer, harvested early, meets those requirements.

Another common specification by the processor is a reasonable size range of disease- and injury-free tubers with no excessively large potatoes. High nitrogen rates greatly increase the proportion of unacceptably large tubers, especially in wet seasons. If in the period of rapid tuber growth there is a short dry spell followed by a return to wet weather, many of these large tubers form external or internal growth cracks and are completely unacceptable to the processor.

Storage potential

Effects of fertilizer rates on storage potential vary considerably from crop to crop. They have little influence on the storage of dried grain, hay or silage but can greatly affect storage losses in potatoes and root crops.

Effects of nitrogenous fertilizers

The main adverse effects on storage quality arise from high levels of nitrogenous fertilizers. Crops such as swedes, which may be left standing in the field for long periods, are most susceptible to excess fertilizer nitrogen. This gives soft, low dry-matter roots which are easily injured. This is particularly serious in some recently introduced varieties which have a naturally low dry-matter content and are susceptible to bacterial or fungal rots introduced through any blemish or injured areas. The rots can spread rapidly in clamps or covered stores. Mangolds and potatoes are similarly affected and injury is unavoidable even with careful handling at harvest and store loading. Crop losses at these stages can be high when fertilizer nitrogen rates are excessive, thus greatly reducing the optimum fertilizer rates for usable or saleable products that have been stored.

The potato crop is particularly susceptible and losses of 2.0–3.0 t/ha through discards on riddling out after storage are common. Figure 14.2 illustrates the effect of increasing the rate of nitrogenous fertilizer on the proportion of damaged tubers of three of the older potato varieties passed over rollers to simulate harvest damage. The percentage of damaged tubers is alarming and is greatest in the variety Majestic which has normally a low

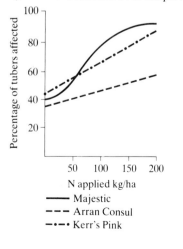

Figure 14.2 The effect of nitrogenous fertilizer rates on the susceptibility of potato tubers to mechanical damage. Data supplied by D.C. Graham and P.C. Harper.

Percentage of tubers affected

N applied kg/ha

—— Majestic
--- Arran Consul
—•—• Kerr's Pink

dry-matter content. Such damage has been reduced by many growers in recent years by modifications to equipment, care at harvest and riddling but some damage is inevitable and will lead to subsequent losses through rotting during storage.

Effects of phosphorus fertilizers

Phosphorus fertilizers used at optimum rates are reputed to improve the storage quality of crops but confirmed evidence is lacking. There is certainly a tendency in most crops for the dry matter content to increase with rate of fertilizer phosphorus and this would give tougher, less disease-prone produce.

Effects of potassium fertilizers

A special example of the benefits of fertilizer potassium on storage quality is its role in the prevention of enzymic browning of potato tubers which renders parts of the tuber black or brown. Although such tubers may be edible they are certainly unacceptable to the housewife. The browning may develop during growth, storage or even during preparation for cooking. Potassium deficiency enhances it and it is more serious when high rates of nitrogen are used. It is greatly reduced or eliminated by the use of optimum rates of potassium fertilizers.

Pests and diseases

Both insufficient and excessive fertilizer rates can increase the susceptibility of growing crops to certain pests and diseases. This is usually the result of a weakening of cell walls which are then easily disrupted by insect attack, wind or mechanical injury from farm implements. All plants deficient in any nutrient are vulnerable. Phosphorus deficient plants are particularly prone to fungal attack. Potassium deficient plants are very susceptible to aphid attack and virus diseases. Nitrogen deficient plants are also attacked preferentially by a wide range of disease organisms. Similarly plants showing trace element deficiency symptoms are subject to fungal infections through lesions. Typical examples are the secondary rots in swedes or sugar beet suffering from boron deficiency.

At the other extreme, excess fertilizer nitrogen weakens the plant by causing the formation of large thin-walled cells, giving easy entry to fungi, pests, pest-borne viruses and bacteria. An example of this is given in Fig. 14.3. The potato crop in question was injured by severe winds and subsequently invaded by grey mould (*Botrytis cinerea*). High levels of nitrogenous fertilizer

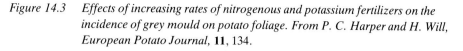

Figure 14.3   *Effects of increasing rates of nitrogenous and potassium fertilizers on the incidence of grey mould on potato foliage. From P. C. Harper and H. Will, European Potato Journal, **11**, 134.*

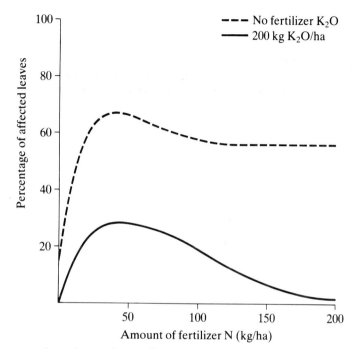

greatly enhanced the disease but increasing rates of potassium fertilizer effectively lessened the incidence.

Appearance, flavour, smell and texture

The so-called organoleptic properties of food – appearance, flavour, smell and texture – are regarded as more important in human foods but animals can also discriminate very well in terms of palatability if given a choice of foods.

It is in these difficult-to-measure properties that problems arise over methods of assessment since they are all judged so subjectively by human beings. Furthermore excellent appearance is no certain guide to flavour – the most attractive looking potato may taste like soap. There are also obvious personal and regional preferences and sometimes consumers are sharply divided in their assessments. The average Scottish housewife would frown upon the soapy texture of potato varieties popular in England, preferring the starchy mealiness of high dry-matter varieties.

Results from tasting panels are commonly inconclusive. Expert panels are essential if there is to be any chance of success in the tests. As an example, a panel of twelve people selected from more than forty as being capable of discrimination, assessed flavour, texture, consistency and appearance of a single variety of potatoes grown with all combinations of three rates of fertilizer nitrogen and three rates of potassium plus a control crop which had received no fertilizer at all. Ten of the twelve panellists agreed that the 'best' potatoes were those grown without fertilizer. There was no agreement about the ranking of the other nine samples.

The most that can generally be established is the existence of 'off-flavours' produced, for example, by excessive rates of fertilizer nitrogen for potatoes. These flavours are associated with increases in the content of certain amides in the tubers. The overwhelming mass of evidence indicates that even specially selected panels cannot generally distinguish between the organoleptic properties of most food crops grown with low or moderate fertilizer applications. It is only when fertilizers are used to excess that differences can be detected in the produce with certainty.

Chapter 15 Fertilizer recommendations, selection and use

Precision in formulating fertilizer recommendations just cannot be achieved. The position in crop production is similar to that in any industry which has to cope with a factor or factors outside the control of the management.

In agriculture the great uncontrollable factor is the weather, which has such a strong influence on nutrient availability to the plant. In the British Isles and northern Europe the climate and especially the day-to-day variations in the weather, compounded by soil variations within a field, make the devising of firm fertilizer recommendations very difficult.

The problem is less severe in regions such as the Corn Belt of the USA where the growing-season climate is much more predictable and there are vast areas of relatively uniform soil. In most maritime regions, however, especially where glaciation and other soil-forming processes have produced variable soil, precise fertilizer recommendations are impossible. For this reason 'blanket recommendations' for a crop must be avoided.

The greatest problem is that of improving the forecasting of nitrogen requirements. Much research is aimed at this but at present soil analysis is of no value and we must fall back on rule-of-thumb methods based on previous cropping, manuring and fertilizing and soil type. The great imponderable factors are the amount of nitrogen that a particular soil will release for a particular crop in a particular season and the amounts of available nitrogen that will be leached during the season. We have no precise methods of allowing for these important factors.

The forecasting of phosphorus, potassium and magnesium requirements is more precise because soil analysis is a useful guide but even for these elements the amounts obtainable by a particular crop from the soil alone are not precisely predictable.

The most that can be done at present is to adjust the type, quantity, time and method of application of fertilizers to maximize their efficiency. General guidance is given below on the principles involved and on the best ways of using the excellent fertilizer recommendations put out by government advisory agencies and reputable fertilizer firms.

Routine fertilizer use

The importance of keeping records

Record keeping is commonly neglected by the hard-pressed farmer. It is tremendously helpful if accurate detailed records of cropping, weather and management data can be kept to assist in decisions on fertilizer type, rate, and method and time of application. Such records are, in fact, indispensable in deciding upon the all-important rates of nitrogenous fertilizer to be applied.

Fertilizer recommendations made by the Agricultural Development and Advisory Services (ADAS) in England and Wales and by the Scottish Agricultural Colleges in Scotland assume that soil conditions are 'satisfactory' for the crops to be grown. If crops are grown in conditions which restrict their growth in any way (poor drainage, pans, trace element deficiencies) the efficiency of the fertilizer will be reduced and the optimum fertilizer rates may be either increased or decreased (Fig. 13.5) but it is certain that yields will suffer.

The following should be carefully checked, monitored and recorded as a routine.

Meteorological records and weather Early access to local meteorological records is essential to permit adjustments to fertilizer nitrogen rates based on winter rainfall and the consequent leaching of nitrates. It is equally important to keep simple records of local weather and especially of unusual events such as short periods of very heavy rain, low temperature or drought. Such events, which do not appear in monthly meteorological records, may help to explain final yield variations.

Cropping history A detailed record of cropping history and crop yields should be kept. This is particularly important in deciding on the rate of nitrogenous fertilizer to be applied for the next crop.

Both the duration and quality of any grass/clover crops should be noted, especially the amount and vigour of clover in the sward.

History of liming, fertilizers and manures   A record of fertilizer and manure use, along with crop yields, is useful in assessing the success of previous policies to assist in deciding rates for the coming crops.

Detailed records of liming and periodic pH checks are essential. The soil pH must be kept well within the optimum range for the crops to be grown (see Fig. 7.3, p. 69) If a good regular liming routine has been established it will be sufficient to have the pH checked by a competent laboratory once every three or four years unless fertilizer use is very intensive, in which case annual checks are desirable.

Applications of slurry or farmyard manure for the current crop should be monitored as accurately as possible and recorded. Both amount and time of application are important in adjusting fertilizer nitrogen rates for the crop. Amounts only are important in adjusting fertilizer phosphorus and potassium rates.

The residual effects of fertilizers and manures on the phosphorus, potassium and magnesium status of the soil are usually reflected in the soil analysis data on which ADAS base their index system which plays a large part in determining their recommendations. Regular soil analysis is essential to the system. Records of analytical results over a period of years will give some indication of fertility trends.

Field boundary changes   The removal of hedges can result in wide differences of chemical fertility in parts of the new field because of the cropping and fertilizer history of the component fields. In such cases pH records are critically important because one part of the field may become acidic enough to cause crop failures while the remainder has a much higher pH.

Cultivations and harvest   A complete record of cultivations and harvest procedures, especially when they have been unavoidably carried out in adverse weather conditions in heavy soils, will help to assess the likelihood of surface compaction or the more insidious unseen sub-surface pans or poor subsoil structure. These conditions can seriously impair the effects of fertilizers on yield. If suspected, they should be checked by the methods described on pages 124–9 of *Soil*, the companion book in this series. They may then be remedied by deep cultivation or subsoiling.

Deficiency diseases   Magnesium, sulphur or trace element deficiencies in previous crops should be recorded so that preventive measures may be taken for the coming crop or corrective measures may be applied without delay if symptoms appear.

The range of NPK recommendations for crops

The broad crop requirements for N, P and K are given in Table 15.1. Around these general requirements have been built tables of fertilizer recommendations showing wide variations according to previous cropping, place in rotation, residual nutrients from previous fertilizers and manures, soil types, abnormal winter and spring rainfall, the use of slurry or farmyard manure for the current crop. In each case the recommendations are based on the best estimates that can be made of the optimum rates of fertilizer nutrients for the crop.

To illustrate the principles involved in devising recommendations two methods are compared below – that used by ADAS and that used by the East of Scotland College of Agriculture for its advisory region.

Table: 15.1   Range of fertilizer recommendations for the main arable crops.

Crop	England and Wales*			South-east Scotland†		
	N	P_2O_5	K_2O	N	P_2O_5	K_2O
Ware potatoes (maincrop)	50–220	100–350	100–350	110–200	110–210	120–300
Sugar beet	25–140	50–100	75–300	—	—	—
Winter wheat	Nil–275	Nil–130	Nil–170	80–185	50–80	50–80
Barley	Nil–160	Nil–130	Nil–170	60–175	50–80	50–80
Swedes	—	—	—	90–110	100–150	90–140
Oil–seed rape	Nil–290	Nil–100	Nil–90	125–260	50–90	50–90

NB   The data in this table do not include adjustments for high or low winter rainfall.

* ADAS recommendations
† Recommendations of the East of Scotland College of Agriculture

Source: Derived from data in ADAS Booklet 2191 and East of Scotland College of Agriculture Bulletin 28.

Table 15.1 gives the range of N, P_2O_5 and K_2O recommendations in kg/ha for the main groups of crops in England and Wales and, separately, in south-east Scotland. The range is very wide and accurate selection within that range is critical.

Adjustments to recommendations

Some adjustments to recommended rates are built in to the published tables as, for example, adjustments for soil analysis in the ADAS tables (the P_2O_5 rate for sugar beet varies from zero to 100 kg/ha according to soil analysis). Other adjustments must be separately applied to the rates given in the main tables – a case in point being the adjustments needed to rates of nitrogenous fertilizers following abnormally wet or abnormally dry winters.

The ADAS index method

P, K and Mg indices   Adjustments to recommendations for P, K and Mg are made by classifying the soils by indices derived from soil analysis. Soils are graded in increasing order of the amount of available nutrient by numbers 0 to 9. Soils with indices of 0 or 1 are deficient and need more of the nutrient in fertilizer form than those with higher indices. For soils of index 2 much smaller effects of fertilizers on yield are expected and none would be expected for soils of index 3 or more. The higher indices (5–9) apply mainly to horticultural soils and crops.

Table 15.2   Recommendations of P_2O_5, K_2O and MgO rates made by ADAS for sugar beet.

Index	P_2O_5 (kg/ha)	K_2O* (kg/ha)	MgO (kg/ha)
0	100	200	100
1	75	100	50
2	50	75	Nil
3	50	75	Nil
over 3	Nil	Nil	Nil

* These recommendations assume that agricultural salt (sodium chloride) has been applied at 400 kg/ha. If not, the recommended K_2O rate is increased by 100 kg/ha or more.

Source: From *Lime and Fertilizer Recommendations No. 1. Arable Crops 1985/6*, Booklet 2191. ADAS, Ministry of Agriculture, Fisheries and Food.

Recommended rates of application at index 3 or more (or for some crops index 2) are simply maintenance dressings intended to replace the phosphorus, potassium or magnesium removed in the crop. Table 15.2 shows the system in practice for sugar beet. The rates of 50 kg P_2O_5/ha and 75 kg K_2O/ha recommended at index 2 and index 3 are maintenance rates. The recommendation of *nil* P_2O_5 at indices of 3 or more implies a conscious decision to run down the very high reserve of phosphorus in the soil.

The success of this index method, based on soil analyses, requires complete confidence in the results of such analyses. Considering the hazards of soil sampling and analysis such confidence may not be justified. It is certainly not justified if inexperienced samplers take inadequate samples from a field. Confidence is reduced still further if even the best soil testing kits are used in do-it-yourself soil analysis without following the maker's instructions implicitly.

The massive reductions in numbers of soil samples analysed by ADAS in recent years along with government insistence on 'farmer sampling' have both tended to erode the basis for this method of forecasting fertilizer requirements, based as it is on the need for reliable and recent soil analysis data.

N index   Unfortunately soil analysis for available nitrogen has little or no value in forecasting the fertilizer requirement of crops. This is recognized by ADAS and they base their nitrogen indices (Tables 15.3 and 15.4) not on soil analysis but on rule-of-thumb estimates of residual nitrogen in the soil from previous crops, manures and fertilizers. The ADAS N index also recognizes the errors inherent in such a method by having only three levels of the index – 0, 1 and 2. The indices are backed up by the results of a very large number of field experiments performed over a period of fifty years and, while they can never be precise, they give a reasonably good guide to the nitrogen requirements of crops.

Essentially the index is determined in mainly arable rotations by considering the likely nitrogen residues from the previous crop only. Residues from crops prior to that are regarded as having no value. If there has been 'permanent' pasture or grass leys of three or more years duration ploughed out or if the last crop was lucerne the indices are modified to allow for enhanced residual values. Account is taken of the nitrogen status of these crops as estimated by fertilizer nitrogen input and sward clover content.

Table 15.3   The ADAS nitrogen index based on the last arable crop grown.

Last crop	Nitrogen index
Any crops receiving large frequent dressings of FYM or slurry	2
Beans	1
Cereals	0
Forage crops removed	0
Forage crops grazed	1
Maize	0
Oil-seed rape	1
Peas	1
Potatoes	1
Sugar beet, tops ploughed in	1
Sugar beet, tops removed	0

Source: From *Lime and Fertilizer Recommendations No.1, Arable Crops 1985/86*, Booklet 2191. ADAS, Ministry of Agriculture Fisheries and Food.

Table: 15.4   The ADAS nitrogen index based on past cropping with lucerne, grass/clover leys and permanent pasture.

Crop	Years since ploughing out				
	0	1	2	3	4
Lucerne	2	2	1	0	0
1–2 year ley, cut	0	0	0	0	0
1–2 year ley, grazed, LN*	0	0	0	0	0
1–2 year ley, grazed, HN†	1	0	0	0	0
Long ley, LN	1	1	0	0	0
Long ley, HN	2	2	1	0	0
Permanent pasture, poor quality matted	0	0	0	0	0
Permanent pasture, average	2	2	1	1	0
Permanent pasture, HN	2	2	2	1	1

* LN = less than 250 kg/ha N per year *or* low clover content
† HN = more than 250 kg/ha N per year *or* high clover content

Source: From *Lime and Fertilizer Recommendations No.1, Arable Crops 1985/86*, Booklet 2191. ADAS, Ministry of Agriculture Fisheries and Food.

Nitrogen indices according to some previous arable crops are given in Table 15.3 and those applied when grassland has been ploughed out are given in Table 15.4. As an example of the system in operation, Table 15.5 shows the adjustments to recommended fertilizer nitrogen rates for the barley crop according to N index with further modifications according to soil type.

Table: 15.5   ADAS recommendations of nitrogen rates (kg N/ha) for winter barley according to N index and broad soil groups.

Soil type	Nitrogen index		
	0	1	2
Sandy soils, shallow soils over chalk or limestone	160	125	75
Other mineral soils	160	100	40
Organic soils*	50	Nil	Nil
Humose soils†	90	45	Nil

* Peats, loamy peats and peaty loams
† Mineral soils containing 6–25% organic matter in the topsoil

Source: From *Lime and Fertilizer Recommendations No.1, Arable Crops 1985/86*, Booklet 2191. ADAS, Ministry of Agriculture Fisheries and Food.

The recommendations are also modified to allow for:
● Farmyard manure and slurry applications.
● Soil types (for potatoes, sugar beet, cereals).
● Winter rainfall (for cereals only).
Other modifications may also be made by the user to allow for other circumstances, for example:
● Heavy rainfall in the early spring.
● Intended use of the crop (malting barley, bread wheat).

Farmyard manure and slurry applications. These allow reductions in the recommended rates of N, P_2O_5 and K_2O. Standard reductions are quoted in the ADAS recommendations. The value of both farmyard manure and slurries is variable and the data in Table 8.1 (see p. 89) may be used to allow for these variations which depend upon the care taken in making, storing, distributing and incorporating the materials. Careful recording at all stages of making, storing and applying the materials is essential for the success of these fine adjustments.

Soil types. By grouping the results of very extensive field experiments ADAS have been able to identify different yield responses to nitrogenous fertilizers on broad groups of soils. This effect can strongly modify the recommendations made according to the N index. This is illustrated in Table 15.5 for winter barley.

Essentially there are 'standard' N rates of 160, 100 and 40 kg/ha for indices 0, 1 and 2 respectively. These apply to most mineral soils. Any modifications are strongly dependent on the organic matter and leaching potential of the soils. Strongly leached, low organic matter soils need more fertilizer nitrogen for a given index and organic soils need less than the standard amounts recommended.

Winter rainfall. Reductions in nitrogen rates are recommended for autumn-sown cereals only if grown in soils with an N index of 1 or 2 and only if the winter rainfall (October–February) has been very low. This is the only case in which it is considered that *mineral* nitrogen residues from the previous crop will not have been leached. It refers mainly to areas in the east of England.

Other factors. The grower may wish to modify the ADAS recommendations to allow for other management factors. These would include the need to restrict grass production in the light of limited stocking, which would involve lower fertilizer inputs. The grower keen to produce malting barley may wish to use less fertilizer nitrogen than ADAS recommend while growers of bread wheat, particularly if they are using a growth retardant, may take a risk by exceeding the recommended rates.

A more difficult decision arises during very wet spring weather when nitrate leaching from freely drained soils can seriously affect requirements of nitrogen fertilizer. It is very difficult to give firm guidance as to the need for top-dressings in such circumstances, but after two or three weeks of heavy spring rain an extra 20–30 kg/ha of fertilizer nitrogen could well be profitable.

The East of Scotland method

This method adopts a different approach from that used by ADAS in that more emphasis is placed on discussions between the farmer and his local adviser. They will modify the basic recommendations in the tables in the light of local knowledge, practical experience and extreme soil types (peats, free draining sands). There are no modifications according to soil type in the actual Tables.

The all-important nitrogen recommendations are based on farming systems (see p 169) and modifications to them are made according to the position of the crop in the rotation. An example of this, for the potato crop, is given in Table 15.6.

Table: 15.6   Effect of farming system and position of crop in rotation on recommended rates of nitrogenous fertilizer for maincrop seed potatoes (East of Scotland).

A. Intensive cash cropping farms – driest areas (700 mm rainfall per year)

Place in rotation	Recommended N rate (kg/ha)
After 3 or more successive cereal crops	165
After other crops	140

B. Arable farms with approximately one-quarter of the cropping area in rotation grass, including a livestock enterprise

After 3 successive cereal crops	155
After 1 or 2 successive cereal crops or other arable crops	130
After 2 or more years grass (unless poor and matted)	100

C. Grass/arable farms with at least one half of the cropping area in rotational grass

After an arable crop	80
After 3 or more years grass (unless poor and matted)	60

NB   The data in this table refer to crops which have received no farmyard manure or slurry.

Source: Adapted from *Fertilizer Recommendations*, Bulletin No. 28, East of Scotland College of Agriculture.

The essential difference between the approach of ADAS and that in the east of Scotland results from the greater importance of grass/clover crops in the farming systems of the east of Scotland as compared with many areas of England. It is, thus, possible in the east of Scotland to base the *main* nitrogen recommendations for arable crops on the nitrogen residues from grass/clover. There is little doubt that the east of Scotland system could be successfully operated in Wales, south-west England and large areas of northern England.

The east of Scotland method also differs from that of ADAS in the smaller emphasis placed on soil analysis for P and K when adjusting rates. Increases in rates of phosphorus and potassium fertilizers are recommended for soils with low analysis (ADAS index 0 or 1) but no emphasis is placed on savings in fertilizers to be made if the analysis is high (ADAS index 2 or more). Specialist soil samplers are employed in Scotland as compared with the emphasis on sampling by the farmer in England and Wales. Also many more samples are analysed annually as compared with similar areas in the south. Thus more accurate and comprehensive data are available in the east of Scotland. Despite this, soil analysis is regarded in this system as having very limited value and justifying only limited adjustments to recommendations.

Modifications to recommendations based on manure applications are similar in the two methods.

In contrast to the ADAS system, however, there are no modifications based on soil type written into the tables in the east of Scotland. The main reason for this is that, in the cool climate with a short growing season, soil type has less influence on the response of crops to fertilizers than in the warmer conditions of the south.

In the east of Scotland method the modifications for winter rainfall place more emphasis on excess rain than do ADAS. Adjustments are made (approximately 25 kg N/ha) for either dry or wet winters using the criterion of 70 mm more or less rainfall than average in October–February.

The accuracy of the recommendations

The two methods of devising recommendations have been used to illustrate the problems involved and also to indicate the somewhat shaky evidence on which they are based. This is no reflection on the authors of the recommendations. They are the best that can be achieved with present knowledge. The problem lies and will continue to lie in the unpredictable nature of the weather and the great variability of soil, particularly in cool, temperate glaciated regions.

The imponderables The potential accuracy of fertilizer recommendations is reduced by four unpredictable and uncontrollable factors:
- Overall growing-season rainfall.
- Day-to-day variations in rainfall.
- Overall growing-season soil and air temperatures.
- Day-to-day variations in temperature.

Between them they have dominant influences on various aspects of fertilizer efficiency:

- Speed and degree of leaching, especially of nitrate and sulphate.
- Rate and amount of mineralization of organic nitrogen, phosphorus and sulphur.
- Rate of plant root growth and ramification within the soil.
- Rate of diffusion and mass flow of fertilizer nutrients within the soil.
- Rate of temporary or permanent fixation of nitrogen, phosphorus or potassium from fertilizers by chemical or biological means.

All these factors strongly affect the ability of the plant to take up nutrients applied in fertilizers. Some of them, such as mineralization, affect the efficiency of the fertilizer by presenting the plant with an alternative nutrient source. Others, such as leaching and fixation, directly reduce the amount of fertilizer nutrient that the plant can obtain. The greatest problem is undoubtedly that of predicting the amount of fertilizer nitrogen required by the crop. This is fundamentally affected by all the factors mentioned above and it is critically important that nitrogen recommendations should be as accurate as possible.

What is apparently insuperable is the time factor. We need to apply a large proportion of the fertilizer requirement during the early part of the season. There is no way of monitoring leaching and fixation sufficiently quickly to apply compensatory fertilizers in mid-season. If vigorous mineralization occurs because of warm moist conditions there is no way of withdrawing fertilizers already applied. There seems to be no alternative but to accept the imponderable nature of the weather and consequent lack of accuracy of recommendations.

Using the tables of recommendations

However the recommendations have been devised they can seldom be simple as so many factors affect the amounts of fertilizer nutrients needed. For this reason simple 'blanket recommendations' for a crop are always suspect. It is essential when using ADAS or similar recommendations to follow very carefully the adjustments, already described, to the recommendations in the main tables.

Fulfilling the recommendations Once the amounts of N, P_2O_5 and K_2O required per hectare have been decided from the tables of recommendations they must be translated into amounts and types of actual fertilizers to be used. Part of this process is to decide whether all the fertilizer should be applied as one dressing in the form of a compound or in split dressings as compound and/or simple fertilizers.

Types and amounts of fertilizer

The selection of fertilizers is best illustrated by taking examples.

EXAMPLE 1 *Choice of solid and liquid compound fertilizers*
The requirements of fertilizer nutrients for a crop of potatoes have been calculated as:

100 kg N/ha
100 kg P_2O_5/ha
150 kg K_2O/ha

It has been decided to apply the fertilizer as a single dressing of a compound fertilizer.
This requires a nutrient ratio of 1 : 1 : 1.5.
There are three suitable fertilizers of that ratio in the sellers' lists, of which two are solids, one with an analysis of 10 : 10 : 15 and the other 13 : 13 : 19.5. Both have the required nutrient ratio but the latter is more concentrated. A liquid fertilizer with the analysis 8 : 8 : 12 (weight/volume) is also available.

The solid fertilizers
Amount of nutrients per bag. The standard bag of fertilizer weighs 50 kg. One hundred kg of 10 : 10 : 15 will supply 10 kg N, 10 kg P_2O_5 and 15 kg K_2O.
Therefore a 50 kg bag will supply half these amounts, i.e. 5 kg N, 5 kg P_2O_5, 7.5 kg K_2O.
Similarly a 50 kg bag of 13 : 13 : 19.5 will supply 6.5 kg N, 6.5 kg P_2O_5, 9.75 kg K_2O.
The number of bags required per hectare. Calculated by dividing the amount of N, P_2O_5 and K_2O required (in this example 100 kg N, 100 kg P_2O_5 and 150 kg K_2O) per hectare by the amount per bag. For the less concentrated fertilizer (10 : 10 : 15), taking the values for N, this would be:

$$\frac{\text{Total requirement}}{\text{Amount per bag}} = \frac{100}{5} = 20 \text{ bags per hectare}$$

The same result is obtained by using the P_2O_5 or K_2O data.

For the more concentrated fertilizer (13 : 13 : 19.5)

$$\frac{100}{6.5} = 15.4 \text{ bags would be needed per hectare.}$$

The number of bags required per acre. May be calculated by dividing bags per hectare by 2.471 (2.5 is sufficiently accurate). Thus 10 bags/ha = 4 bags/acre and the number of bags of 10 : 10 : 15 required per acre in this example is 8; 6.16 bags of 13 : 13 : 19.5 would be needed per acre.

The liquid fertilizer Before calculating the amount of the 8 : 8 : 12 liquid fertilizer required per hectare it is necessary to confirm that the specification of 8 : 8 : 12 refers to the weight of nutrients in a given volume of fertilizer (weight/vol). If this is the case an 8 : 8 : 12 fertilizer will contain 8 kg N, 8 kg P_2O_5 and 12 kg K_2O per 100 litres. The seller of liquid fertilizers is legally allowed to quote the composition of the fertilizer either in this way or on a weight/weight basis – that is in terms of kg of N, P_2O_5 or K_2O per 100 kg (or alternatively per tonne) of fertilizer. This makes a considerable difference to the amount of N, P_2O_5 or K_2O that would be supplied by a given volume of liquid. The difference depends on the density of the liquid and this varies from one formulation to another but it will certainly be more than 1 and is commonly about 1.2.

On this assumption a fertilizer quoted as 25 per cent N weight/volume would be equivalent to the same fertilizer quoted as only 20.8 per cent N on a weight/weight basis.

In this example the liquid fertilizer is 8 : 8 : 12 (weight/vol).

This will contain 8 kg N, 8 kg P_2O_5 and 12 kg K_2O per 100 litres. The amount required per hectare would be

$$\frac{100}{8} \times 100 = 1\,250 \text{ litres}$$

This example shows the simplest problem in fertilizer selection in that all the fertilizer was to be applied as a single dressing and there were three similar compounds of exactly the correct ratio available. In this case the decision as to which of the three to adopt hinges on comparative prices, including any reductions from the list prices, ease of application and suitability for the farm management system.

EXAMPLE 2   *Fertilizers for split dressings*
Winter barley is the crop. From the tables of recommendations

125 kg N/ha, 50 kg P_2O_5/ha and 50 kg K_2O/ha

are required. Because of an abnormally wet summer in the previous year and the fact that the two preceding crops were cereals, a small proportion (25 kg/ha) of the nitrogen is to be applied at sowing along with all the phosphorus and potassium. The remainder of the nitrogen (100 kg/ha) is to be applied in spring shortly before the start of growth, thus:

Autumn: 25 kg N, 50 kg P_2O_5, 50 kg K_2O
Spring: 100 kg N

Only one compound fertilizer is available in the lists with a suitable ratio for the autumn application. It is a 10 : 20 : 20 fertilizer (ratio 1 : 2 : 2). Each 50 kg bag would, therefore, contain 5 kg N, 5 kg P_2O_5, 10 kg K_2O. The requirement per hectare would be $25/5$ = 5 bags/ha. For the nitrogen application in the spring there is a choice of several fertilizers. They are:
   A Ammonium nitrate (solid), 34.5 per cent N
   B Urea (solid), 44 per cent N
   C Ammonium nitrate/calcium carbonate (solid), 26 per cent N
   D Ammonium nitrate/urea (liquid), 25 per cent N (weight/vol).
   As an example of the calculation of the number of bags per hectare, fertilizer A contains

$$\frac{34.5}{2} = 17.25 \text{ kg N/bag}$$

Therefore $\dfrac{100}{17.25}$ = 5.8 bags/ha will supply the necessary 100 kg N/ha.

Similar calculations show that 4.5 bags of fertilizer B or 7.7 bags of fertilizer C would be equivalent to the 5.8 bags of fertilizer A, while

$$\frac{100}{25} \times 100 = 400 \text{ litres/ha of fertilizer D would be required.}$$

The choice between these fertilizers must then be made in terms of price, convenience and an estimate of comparative effectiveness.

More difficult cases   In Examples 1 and 2 fertilizers were available to fit precisely the recommended rates. Despite the wide range of concentrations and ratios of fertilizers available it is not always possible to find the precise formulation required.

In such cases the first priority is to select a fertilizer and rate of application which will give exactly the required amount of *nitrogen*. It should also supply approximately the recommended amounts of P_2O_5 and K_2O but this is rather less important. If the soil in question has low P or K indices (0 or 1) it is preferable to select a fertilizer with a slightly higher ratio of P and K to N than that desired. If the P and K indices are high, especially with indices of 3 or more, a lower ratio of P and K to N is then acceptable because the recommended phosphorus and potassium rates are for maintenance only and little or no yield response is expected in the crop.

EXAMPLE 3   *No fertilizer available to fit the recommended rates precisely, soil P and K indices high*
Soil P index 3   Soil K index 3
Recommended rates: 100 kg N/ha, 60 kg P_2O_5/ha, 80 kg K_2O/ha
Ratio: 1 : 0.6 : 0.8

Three fertilizers are available with ratios reasonably similar to the required ratio:

A 10 :   5 :   5 (ratio 1 : 0.5 : 0.5 *or* 2 : 1 : 1)
B 10 :   8 : 10 (ratio 1 : 0.8 : 1)
C 10 : 10 : 10 (ratio 1 : 1 : 1).

Any of these fertilizers applied at 20 bags/ha will supply the recommended 100 kg/ha of nitrogen.

Total amounts of nutrients supplied by the three fertilizers in kg/ha would then be:

	N	P_2O_5	K_2O
Fertilizer A	100	50	50
Fertilizer B	100	80	100
Fertilizer C	100	100	100

Thus, fertilizers B and C would both exceed the recommended rates of P_2O_5 and K_2O. They would be likely also to cost more than fertilizer A. In this soil with high P and K indices the phosphorus and potassium supplied by fertilizers B and C would exceed maintenance requirements whereas A would not quite meet them. Fertilizer A would be most acceptable. This decision implies a slight run down in the high levels of available P and K.

EXAMPLE 4
No fertilizer available to fit the recommended rates precisely, soil P and K indices low
Soil P index 0   Soil K index 0
Recommended rates: 80 kg N/ha, 110 kg P_2O_5/ha, 140 kg K_2O/ha
Ratio: 1 : 1.4 : 1.8
Three fertilizers are available with ratios reasonably similar to that required:

A 12 : 12 : 16 (ratio 1 : 1 : 1.3)
B 12 : 18 : 24 (ratio 1 : 1.5 : 2)
C 12 : 16 : 20 (ratio 1 : 1.3 : 1.7).

Any of these fertilizers applied at 13.3 bags/ha will supply the recommended 80 kgN/ha.
   Total amounts of nutrients then supplied by the three fertilizers in kg/ha would then be:

	N	P_2O_5	K_2O
Fertilizer A	80	80	107
Fertilizer B	80	120	160
Fertilizer C	80	107	133

In this case fertilizer B should be selected and applied at 13.3 bags/ha. The other fertilizers applied at the same rate would supply less than the recommended amounts of phosphorus and potassium and this would be risky on this soil which is deficient in both.

The use of simple fertilizers

In Examples 1–4 it would have been possible to use simple N, P and K fertilizers to fulfil the recommendations accurately. This was what happened before the advent of compound fertilizers and there could well be a case for returning to simple fertilizers. The nutrients in them are usually cheaper unit for unit than those in compound fertilizers because of the lower costs of manufacture.

EXAMPLE 5
Simple fertilizers
Recommended rates as in Example 3, viz:

100 kg N/ha, 60 kg P_2O_5/ha, 80 kg K_2O/ha
Ratio: not required in this calculation.
It has been decided to use:

Urea (45 per cent N)
Triple superphosphate (45 per cent P_2O_5)
Muriate of potash (60 per cent K_2O).

The number of 50 kg bags of each to supply the required amounts of N, P_2O_5 and K_2O would be:

Urea	$\dfrac{100}{22.5}$	= 4.4
Triple superphosphate	$\dfrac{60}{22.5}$	= 2.7
Muriate of potash	$\dfrac{80}{30}$	= 2.7

Choice of suitable fertilizers

An important part of the process of fertilizer selection is to ensure that the most effective material is chosen for the circumstances. In some soils, for example, it may be preferable to use urea-based in preference to ammonium nitrate-based nitrogen.

Nitrogen fertilizers It is not correct to assume, as is so often quoted, that nitrate nitrogen is more easily available to the plant than ammonium nitrogen and that ammonium needs to be converted to nitrate before the plant can use it.

Some crops, for example potatoes, appear to prefer ammonium nitrogen but this may be simply a reflection of the soil-acidifying effects of ammonium nitrogen which will be beneficial for potato crops growing in soils of pH values above the optimum range.

In fact ammonium (NH_4^+) and nitrate (NO_3^-) are absorbed with approximately equal ease by plants. Once within the plant NO_3^- must be converted to NH_4^+ before it goes on to form amino-acids and proteins. This process requires energy and thus in theory the plant will use NH_4^+ more efficiently than NO_3^-.

In practice the comparative effectiveness of nitrogenous fertilizers is controlled more by what happens outside the plant than inside, the main factors being nitrate leaching, denitrification and loss of ammonia from the soil by volatilization.

The following general rules are useful in selecting nitrogenous fertilizers:

- On soils where leaching is not a serious problem and the soil pH is in the range of 5.0–6.0 there is little to chose between ammonium, urea or nitrate nitrogen in their effects on yield.
- In high rainfall areas where leaching is a problem, for soils in the pH range 5.0–6.0, especially if sandy and free draining, a high ratio of ammonium plus urea/nitrate nitrogen is preferable.

● For soils in the pH range 6.0–7.0 ammonium or urea nitrogen also has a slight advantage over nitrate.
● On soils in dry areas, where leaching is not a problem and the soil pH is greater than 7.0, nitrate nitrogen is slightly superior to ammonium or urea nitrogen, especially for top dressings.

Unfortunately, in the British Isles, solid compound fertilizers containing exclusively ammonium nitrogen, ammonium plus urea, or a high ratio of ammonium plus urea to nitrate are not easy to come by. The farmer who opts for ammonium nitrogen only may well be frustrated but prilled urea, urea in liquids and liquid fertilizers containing high ratios of urea plus ammonium to nitrate are widely obtainable.

Phosphorus fertilizers There is little to choose between the various types of simple water-soluble fertilizers – superphosphate, triple superphosphate and the water-soluble ammonium phosphates widely used in compound fertilizers.

The choice between water-soluble phosphates, mixtures of soluble and insoluble materials or solely water-insoluble phosphates is important in many circumstances and may affect fertilizer costs considerably.

● For all arable crops grown on soils of low P index (0 or 1) exclusively water-soluble phosphates or fertilizers with a high ratio (4 : 1 or more) of water soluble : water insoluble phosphates should be used because of the need for readily available phosphorus.
● If the available phosphorus in the soil is high or very high (P index 2 or more), and the phosphorus fertilizer is being used simply to replace that which will be removed by the crop, consideration should be given to using a fertilizer with a moderate ratio of water-soluble : water-insoluble P_2O_5 (2 : 1, 1 : 1) especially for crops with low phosphorus requirements such as sugar beet and cereals. Such fertilizers should be cheaper than exclusively water-soluble materials. The richer the soil the less important is the choice of phosphate fertilizer and the cheapest form may be used except for rapid growing market garden crops.
● On grassland in high rainfall areas (850 mm or more per year) and especially on acidic soils (pH lower than 5.7) a simple

water-insoluble fertilizer, or alternatively one with a low ratio of water soluble : water insoluble phosphorus (1 : 2) should be used.

● The choice of water-insoluble phosphate fertilizers is now very restricted in the British Isles. If they are obtainable, basic slags are generally rather more effective than ground mineral phosphates.

● Whichever water-insoluble phosphates are used, they should have a high solubility in the solvent specified by the EEC (Table 9.3, p. 119). *Soft* ground mineral phosphates should be preferred to hard ones.

Potassium fertilizers   The choice usually lies between potassium chloride (muriate of potash) and potassium sulphate. These are the potassium salts normally incorporated in compound fertilizers. In terms of crop yield and quality there is no effective difference between the two for cereals, grass, swedes and sugar beet. For potatoes, grown for processing or 'seed', and for high-value vegetable crops the sulphate form is preferable. It can also make a contribution to the sulphur requirement of crops and is therefore useful for high sulphur-demand crops such as kale and oil-seed rape.

Comparative prices

Simple fertilizers   The price of simple fertilizers must be compared *not* on the cost per tonne of fertilizer but on the cost per unit of N, P_2O_5 or K_2O contained therein.

The most useful unit to use, now that fertilizers are sold in metric terms, is the kilogram. This makes the calculation of unit costs very simple.

EXAMPLE: Fertilizer A contains 34.5 per cent N and costs £135 per tonne.

Fertilizer B contains 26 per cent N and costs £108 per tonne.

Fertilizer C contains 45 per cent N and costs £163 per tonne.

Fertilizer A
  Cost per tonne = £135
  kg N/tonne   = 34.5 × 10 = 345
  Cost per kg N = $\dfrac{£135}{345}$ = 39.1p

Fertilizer B

$$\text{Cost per tonne} = £108$$
$$\text{kg N/tonne} = 26 \times 10 = 260$$
$$\text{Cost per kg N} = \frac{£108}{260} = 41.5\text{p}$$

Fertilizer C

$$\text{Cost per tonne} = £163$$
$$\text{kg N/tonne} = 45 \times 10 = 450$$
$$\text{Cost per kg N} = \frac{£163}{450} = 36.2\text{p}$$

Thus fertilizer C, which costs most per tonne, has the lowest price per kg N. The three 'costs per kg N' are directly comparable. Similar calculations may be made for simple phosphorus and potassium fertilizers using cost per kg of P_2O_5 and K_2O respectively.

Compound fertilizers The comparative costs of compound fertilizers of *identical* nutrient ratios can be calculated by dividing the cost per tonne by the nitrogen content in kg/tonne (in precisely the same way as for simple fertilizers).

Note that the 'notional comparative costs' calculated in this way will not correspond with the costs per kg of nutrient in simple fertilizers but will always be higher because they account for N plus P_2O_5 plus K_2O costs and not just one of the three.

EXAMPLE   Fertilizer 1

Analysis 20 : 10 : 10
Cost per tonne = £140
kg N per tonne = 200
Notional comparative cost = 70p

Fertilizer 2

Analysis 24 : 12 : 12
Cost per tonne = £160
kg N per tonne = 240
Notional comparative cost = 66.7p

Therefore fertilizer 2 will give the same amount of nutrients as fertilizer 1 at a lower cost.

The comparison can be converted into more useful terms:

100 standard bags of Fertilizer 1 are worth the same as $100 \times \dfrac{66.7}{70} = 95.3$ bags of Fertilizer 2.

This method of calculation may be used to give a rough but sufficiently accurate comparison of compounds with similar but not identical ratios. For example it would be justifiable to compare a 1 : 1 : 1.5 ratio fertilizer, in this way, with a 1 : 1 : 1.6 or a 1 : 0.9 : 1.5 as the ratios are close but for comparing a 1 : 1 : 1.5 with a 1 : 1 : 2 or a 1.5 : 1 : 1.5 as the differences between the ratios are too wide.

In practice fertilizers with widely differing ratios are seldom suitable for the same job and the need to compare their costs seldom arises.

Simple versus compound fertilizers The costs of applying equivalent amounts of N, P_2O_5 and K_2O in the form of three separate simple fertilizers or as a compound may be compared by applying the 'cost per kg' of the simple fertilizers as follows:

Nutrient requirements: 100 kg N/ha; 50 kg P_2O_5/ha; 150 kg K_2O/ha. Most suitable compound fertilizer: 12 : 6 : 18
Cost: £126 per tonne.

Cost of suitable simple fertilizers:
Nitrogen: N content, 40%; Cost, £176 per tonne
Phosphorus: P_2O_5 content, 45.5%; Cost, £191 per tonne
Potassium: K_2O content, 60%; Cost, £120 per tonne

Compound fertilizer required/ha =   0.83t; Cost £104.6

Simple N fertilizer required/ha =   0.25t; Cost £44.0

Simple P_2O_5 fertilizer required/ha = 0.11t; Cost £21.0

Simple K_2O fertilizer required/ha = 0.25t; Cost £30.0

Sum of costs of simple fertilizers =   £95.0

Therefore the cost of simple fertilizers (£95) is £9.6 per ha less than that of the compound (£104.6). In the example taken this represents a 10 per cent saving on the costs of using the compound fertilizer. Such savings would need to be set against the inconvenience of buying, storing and applying three separate simple fertilizers.

Convenience
Although suitability and price are major factors in fertilizer selection convenience is also important.

Many farmers have a good relationship with a particular merchant and may be offered not only a very good price but also helpful advice, delivery and even spreading exactly when required and also services such as soil analysis. Only the farmer can assess the value to him of such help.

Farmers with good storage facilities may find it convenient to take advantage of early delivery rebates, offered by fertilizer firms to ease their own storage problems, provided that they can afford to pay early.

Another major convenience comes from the concentration of fertilizers. Fifteen standard bags of a simple fertilizer containing 20 per cent N are equivalent to only ten bags of a 30 per cent N product. This involves much extra labour if the less concentrated fertilizer is used. If 'big bags' are used the difference in labour requirements will not be so great. The use of big bags, however, implies capital expenditure on handling equipment which may not be justified on small farm units.

Depending on the farming system and staffing it may be convenient to have all the fertilizer delivered and spread by contractors, as either liquids or solids.

It is difficult to assess the value of these factors but they must be taken into account when selecting fertilizers.

Time of application
The simplest approach is to apply all the fertilizer nutrients needed by the crop in one dressing of compound fertilizer at or about sowing time. This saves extra runs over the field and reduces problems of damage by fertilizer distributors to a standing crop during top dressing. It does not, however, always lead to the most efficient use of fertilizer nutrients by the plant. Many crops make little growth and take up only small amounts of nutrients in the first four or five weeks after sowing. During this period losses of seedbed fertilizer by leaching or fixation are bound to occur.

Nitrogen is most affected and thus the timing of phosphorus and potassium applications is not so critical as that of nitrogen applications.

Nitrogenous fertilizers
Autumn applications

● Autumn application of the whole crop requirement of nitrogen is very inefficient and should never be undertaken.

● If straw is being ploughed in an extra autumn application of 10 kg N/tonne of straw should be made. This will speed up the microbial decomposition of the straw and with autumn-sown cereals will prevent a shortage of nitrogen for the crop during the autumn period.

● Following an excessively wet summer, especially on nitrogen deficient soils (N index 0), a small autumn dressing of 20–30 kg N/ha is desirable for winter barley or winter wheat.

● For other autumn-sown cereal crops no autumn application of nitrogen should be necessary.

● For autumn-sown oil-seed rape, provided it is sown early enough to make good autumn growth, approximately a quarter (40–60 kg N/ha) of the total nitrogen requirement should be applied at sowing. Higher autumn applications enhance the risk of frost damage.

Spring applications for cereals

● Spring top dressings for winter barley or winter wheat should be kept flexible. Extra nitrogen may be needed after a wet winter, less after a dry winter.

● Allowance may even be made for leaching losses in a wet spring season by use of an extra late top dressing (April–May) of 20–30 kg N/ha for winter wheat or feeding barley but *not* for malting barley.

● For winter wheat, spring top dressings should be varied according to the nitrogen status of the soil. On nitrogen deficient soils (index 0) the recommended fertilizer nitrogen should be split into two separate applications, one in March and the other in late April or early May. On soils with Nitrogen index 1 or 2 all the nitrogen required may be applied in one top dressing during April.

● In the special case of bread wheat improved gluten content may be obtained by a delayed extra top dressing of 20–30 kg N/ha applied in late May.

● Winter barley for malting should receive the whole of its spring requirements of nitrogen fertilizer at or about the onset of spring growth (February–March).

● Winter barley for feeding is better left until late March/early April for top dressing.

● For spring barley, especially for malting, all the fertilizer nitrogen required should usually be applied and incorporated before sowing. In wetter areas on nitrogen-rich soils the application may be delayed until a week or two weeks after emergence.

Spring applications for other arable crops

● For potatoes and sugar beet normal practice is to apply all the fertilizer nitrogen at sowing. Considering the very slow early-season growth of these crops it might be an advantage to apply a half to two-thirds of the total nitrogen at sowing and the remainder two to three weeks after emergence.

Grassland

● It is very inefficient to apply the whole nitrogen requirement in one spring dressing.

● For efficient use, especially in high nitrogen-input systems, it is essential to use split dressings, usually at or immediately before the onset of spring growth and then after each grazing or each cut of grass. Rates of application must be gauged to the expected yield from the following cut. The highest rate should be given for the first cut of grass.

Phosphorus and potassium fertilizers In general the time of application of phosphorus and potassium fertilizers is much less critical than that for nitrogenous fertilizers.

● On acutely deficient soils it is a good principle to apply water-soluble phosphate fertilizers immediately after sowing or a week or two earlier. If applied earlier still some of the advantages of water-solubility are lost through fixation.

● Because of their slow action, water-insoluble fertilizers used for arable crops should always be applied during the autumn of the previous year.

● Autumn applications of water-soluble phosphorus and potassium to autumn-sown crops should be made if the index is low (0 or 1). Otherwise they should be applied in the spring with the first top dressing of nitrogen.

● The higher the soil P and K indices the less important is the time of application of phosphorus or potassium fertilizers. On soils with P or K indices of 2 or more, they may be applied in the autumn preceding the cropping year, if convenient.

● Seedbed applications of phosphorus and potassium fertilizers should preferably be made two or three weeks before sowing to reduce seedling damage.

● Early season applications of potassium fertilizers to grazed grassland are best avoided, especially if the grass is intensively fertilized, because of the increased risk of hypomagnesaemia in the grazing animal. In these circumstances any potassium required is best applied in mid season but the return of potassium to the soil from animal excreta is great and the K index is likely to be high.

Intensively fertilized grass, grown for conservation, with 250–400 kg N/ha should be given phosphorus and potassium along with nitrogenous fertilizers at the start of spring growth and also after each of the first two cuts of grass. If it is more convenient the first application of P and K may be given in the previous autumn.

Sulphur fertilizers   The choice of compound fertilizers containing sulphur is extremely limited.

Only compound fertilizers containing a proportion of ammonium sulphate or potassium sulphate are extensively obtainable as water-soluble sulphur sources. In low pollution areas they should be chosen in preference to sulphur-free fertilizers, especially for crops like kale, oil-seed rape and turnips, which have high sulphur requirements.

Gypsum may be used to supply sulphur on a slightly longer-term basis. It may be applied by fertilizer distributor at 400–500 kg/ha with the object of boosting available sulphur supplies for 2–3 years.

Magnesium fertilizers   Compound fertilizers containing any appreciable amount of magnesium are difficult to obtain.

In cases where magnesium deficiency symptoms occur, especially on potatoes or sugar beet, the crop should be sprayed with a solution of Epsom salt (4 kg/ha in 200 litres of water). In less urgent cases, but with a soil magnesium index of 0 or 1, 200–400 kg of kieserite ($MgSO_4H_2O$) per hectare should be applied for demanding crops such as sugar beet, potatoes, kale, oil-seed rape and turnips.

In all areas of acidic soils where available soil magnesium is low, magnesian limestone should be used as a routine when liming is needed. A standard rate of 5 t/ha will be required every 3–4 years, and more frequently if the soil is intensively cropped.

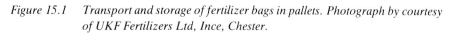

Figure 15.1   Transport and storage of fertilizer bags in pallets. Photograph by courtesy of UKF Fertilizers Ltd, Ince, Chester.

Methods of handling and application

Solid fertilizers

Storage and handling Most solid fertilizers are now sold in granular form in strong polythene bags holding 50 kg. They are easily handled and stored in this form. It is preferable to store them in a dry building but they may be stored outdoors so long as they remain undamaged, if covered by a strong waterproof sheet. If possible, they should be stored in piles five or six high on a raised platform so that when required they do not need to be lifted but may be lowered to the vehicles used to transport or spread the fertilizer. On small farms, using 20–30 tonnes of fertilizer each year, manual handling, though tiring, is probably more efficient than any attempt to mechanize the process.

There are recent developments involving mechanized handling that can be used to advantage on large units.

Pallets (Fig. 15.1) can be used for delivering and handling fertilizer in standard bags, each pallet holding 20–30 bags (1–1.5 t). The pallets can be transported on the farm by fork-lift trucks or by using a special attachment on a tractor front-end loader. Capital costs of equipment are high but, on large farms, there are great advantages in both time and human energy saved, mostly at receipt and during stacking.

Large bags of fertilizer (Fig. 15.2) can also be obtained, containing 500 kg – 1t. The bags consist of a strong woven-plastic cover inside which the fertilizer is sealed in polythene. They may be disposable or returnable. Large bags must be handled mechanically and are most useful where large amounts of fertilizer need to be spread in a very short space of time at high rates.

They can be stacked, no more than two high, and may be handled by hook or crane attachments to heavy-duty fork-lift trucks or tractor front-end loaders. The non-returnable type of bag is simply lifted directly over a hopper-spreader and emptied into it by slitting the bottom of the bag. Returnable bags are designed with an outlet spout for feeding the fertilizer into a distributor.

Bulk delivery and spreading of unbagged fertilizers is done by some contractors and should be left to them. The equipment used is very heavy and soil damage may occur in wet conditions from wheel tracking.

Bulk storage on the farm of unbagged fertilizers is feasible in countries where humidity is consistently low but should be attempted only for short periods in the British Isles.

Figure 15.2   *The use of 'big bags' of fertilizer. Photograph by courtesy of Norsk Hydro Fertilizers, Ipswich, Suffolk.*

Broadcast spreading Most solid fertilizers are now spread by broadcasting, the aim being to distribute the material as uniformly as possible over the soil surface. The fertilizer should then be incorporated in the soil as soon as possible, by harrowing, to avoid volatilization losses of ammonia and to assist uniform distribution within the surface soil.

Broadcasting spreaders The earliest mechanical spreaders consisted of a long trough from which the fertilizer trickled through regularly spaced spouts on to a series of revolving discs which sprayed out the material. Metering was crude and depended on careful observation and control by a second operator in addition to the driver.

Modern spreaders have accurate metering devices using rollers which can be adjusted to give application rates as low as 100 kg/ha or as high as 1 t/ha irrespective of the forward speed of the machine.

Pneumatic spreaders (Fig. 15.3) are the natural successors to the early machines. The fertilizer is fed from a hopper to a metering device and is then blown by a high-speed flow of air through tubes on to spreading plates placed at intervals along a boom about 10 metres wide. The essential characteristic of pneumatic spreaders is to give uniform distribution of the fertilizer to the exact width of the boom but not beyond this. They are, therefore, particularly useful for crops grown on tramline systems. It is essential with these machines to avoid overlapping, which would give double the intended rate, or gapping, in which no fertilizer at all is given. For this purpose most machines are fitted with bout markers (a bout is a single run across a field) which deposit a line of foam or some coloured material to guide the driver for the next bout (Fig. 15.4).

Oscillating spout and spinning disc spreaders rely on overlapping of fertilizer spread from one bout to the next to get even distribution. Thus mistakes in bout width are not nearly so obvious as with pneumatic spreaders, but they are none the less important. Provided that care is taken over bout widths, adjustment of rate mechanisms and maintenance, these broadcasters give a reasonably uniform distribution over well-tilled soil surfaces. Both types of machine depend upon the metering of the fertilizer from a hopper. This is done by adjusting the size of a hole at the bottom of the hopper according to the type of particle (granule or prill) and the required rate of application. The mechanisms below the hopper are designed to throw the fertilizer behind and to the side of the machine.

Figure 15.3   Broadcasting fertilizer using a pneumatic spreader. Photograph by courtesy of A. C. Bamlett Ltd, Thirsk, North Yorkshire.

Figure 15.4   A bout marker in action. Photograph by courtesy of Saltney Engineering Ltd, Bishops Stortford, Herts.

The spinning disc machines (Fig 15.5) throw the fertilizer from revolving discs which may be slightly cupped or may have radial vanes. The oscillating spout machines (Fig. 15.6) deliver the fertilizer from a large spout, the speed of oscillation of which can be adjusted to affect the bout width.

Figure 15.5   Broadcasting fertilizer using a spinning disc machine. Photograph by courtesy of Lely Import Ltd, St Neots, Huntingdon, Cambs.

Both machines give a roughly triangular distribution pattern when in motion and bout widths are recommended by the manufacturers to give an overlap of the extreme edges of that pattern. If well managed these machines deliver half as much fertilizer in the overlap as in the main zone behind the machine and the next bout ensures uniform treatment of the area.

Figure 15.6   Broadcasting fertilizer using an oscillating spout machine. Photograph by courtesy of Vicon Ltd, Ipswich, Suffolk.

Placement   As an alternative to broadcasting the fertilizer may be placed, by special machines, in close proximity to the seed at the time of sowing row crops.

Combine drilling of cereal seed and fertilizer is an example of this technique. Contact placement, in which seed and fertilizer are fed down the same spout, should be avoided because of the risk of salt effects and damage to germinating seedlings. The most successful combine drills place the fertilizer slightly below (4–6 cm) and to the side (4–6 cm) of the seed. This reduces the risk of salt damage while concentrating the fertilizer where it is most needed in the early stages of growth.

Placement of fertilizers had major advantages in the British Isles when many soils were deficient in phosphorus and when modest rates of fertilizer application kept the risk of damage to seedlings small. The advantages have diminished and the disadvantages have increased as soils have become more fertile while the concentration of fertilizers and their rate of application has increased.

The following general rules apply:
- Placement is least effective in fertile soils.
- It is most effective for phosphate fertilizers used in phosphate deficient soils.
- It is useful for widely spaced plants with short growing periods.

In these cases rates of application may be reduced if fertilizer is placed instead of being broadcast.

Liquid fertilizers

Storage Liquid fertilizers are usually delivered to the farm in bulk tankers. They may then be applied immediately or stored in bulk tanks on the farm to ensure immediate accessibility. The storage tanks are usually made of mild steel and are subject to corrosion by simple nitrogenous liquid fertilizers. One of the cheapest anti-corrosion measures is to fill new tanks with a liquid compound fertilizer rich in phosphorus. A corrosion-resistant lining will then be formed. Alternatively, plastic coatings may be applied to the inside of the tank before filling.

Storage during spring and summer presents few difficulties but there may be problems in winter because of the reduced solubility of many fertilizer salts at low temperatures. The problems are mitigated to some extent by the time taken for the very large bulk of liquid in the tanks to cool down. None the less, during prolonged periods of very cold weather some precipitation will occur. If precipitates do form they are very difficult to re-dissolve because they settle out in a solid layer. Manufacturers naturally try to avoid formulations that will give precipitates. Among the most difficult formulations to store are those which include potassium salts and ammonium nitrate which react together to form potassium nitrate, sparingly soluble at low temperatures.

Spreaders Liquid fertilizers are best sprayed on to bare soil and immediately harrowed in. The liquid is pumped from a tank through a long boom to a series of spray nozzles (Fig. 15.7).

Provided that care is taken to maintain the jets, to use the pressure recommended by the manufacturer and to match the bouts, distribution should be very even. Jet sizes can be adjusted to give different application rates.

Figure 15.7   Spreading liquid fertilizer using a long boom. Photograph by courtesy of J. W. Chafer Ltd, Doncaster, South Yorkshire.

Spray applications to standing crops, especially if the droplets are very small, can cause serious scorching of the leaves. This can be partly overcome by using very large droplet sprays which tend to drop rapidly from vegetation to soil. There is, however, inevitably some collection of liquid, for example between stem and leaf and on some hairy-leaved plants, and this will cause some scorch.

There are also dribble-bar attachments by which the liquid is dribbled through flexible plastic tubes to soil level. The tubes may be spaced to avoid contaminating row crops. Placement of liquid fertilizers may be achieved by using flexible tubes, behind coulters, designed to deposit a stream of liquid under the soil surface.

Injectors for gaseous ammonia This very toxic material requires special handling by trained contractors equipped with protective clothing and safety devices. It is injected from cylinders, under pressure, some 10–15 cm below the surface of the soil through plastic tubes placed behind special coulters. Behind the coulters is a device for closing the injection slit to avoid or reduce the loss of ammonia to the atmosphere. The problems of using liquefied anhydrous ammonia are discussed on page 128.

Urgent fertilizer use

By careful management the need for urgent use of fertilizers to supply phosphorus, potassium, sulphur and magnesium should be avoidable.

Nitrogenous fertilizers

The most commonly needed urgent treatment is that of nitrogenous fertilizer. Any deficiency of nitrogen in a crop shows up very rapidly in the appearance of the plants. They become successively pale green, yellow-green and yellow. Usually older leaves are affected first, turning yellow and dying from the tip. The essential feature is an overall yellowing of the leaf, veins and all, brought about by lack of chlorophyll. These symptoms are fairly reliable indicators of nitrogen deficiency but do not indicate the cause.

Table 15.7   Diagnosis of the causes of nitrogen deficiency in crops.

Diagnostic features* in soil, plant and climate	Cause of deficiency	Likely response to nitrogenous fertilizers or manures
Very wet weather, high annual rainfall, sandy or gravelly soil texture, low soil organic matter.	Leaching.	Vigorous response but nitrates in applied fertilizers subject to rapid leaching.
Very dry weather, soil visibly dry.	Drought.	Response only if drought breaks or irrigation can be used. In either case the nitrogen fertilizers applied earlier become effective.

Strongly acidic soil. Stubby scorched roots.	Toxicity of manganese and aluminium, calcium deficiency, lack of mineralization of organic soil nitrogen.	Response only if the acidity problem is resolved by liming.
Straw or other high carbon/nitrogen materials present in excess.	High C/N ratio. Bacteria temporarily immobilizing available nitrogen.	Vigorous response.
Waterlogged soil conditions.	Poor drainage, surface or subsoil pans.	Slight response but drainage problem must be resolved.
Lesions on lower stem or roots (specialist should be called).	Damage by nematodes or specific fungi.	Commonly a limited response but this is restricted because of irreversible damage to the plant.
Cold, wet conditions at the time of seedling emergence.	Temporary failure of root penetration and nutrient absorption by plant.	Usually no need for additional fertilizer nitrogen. In fact this could provide excess nitrogen once the weather improves.
None of the above apply but leaves yellowing especially at early growth stages.	Inadequate or badly timed fertilizer application.	Vigorous response.

* Whatever the *cause* of nitrogen deficiency in the plant, symptoms on the leaves will be similar, viz: pale green, yellow-green or yellow leaves. Usually the older leaves are affected first, turning yellow and dying from the tip.

Early and correct diagnosis of nitrogen deficiency followed by urgent application of nitrogenous fertilizer can be very rewarding but it is absolutely essential to establish what has caused the deficiency. Crop response is strongly dependent upon a correct assessment of the cause and much money and effort can be wasted through faulty assessment. Table 15.7 summarizes the causes of nitrogen deficiency, how to identify them, and also gives the likely response of the crop to urgently-applied nitrogenous fertilizer.

Figure 15.8   *Diagnosis of mineral deficiencies in plants.*

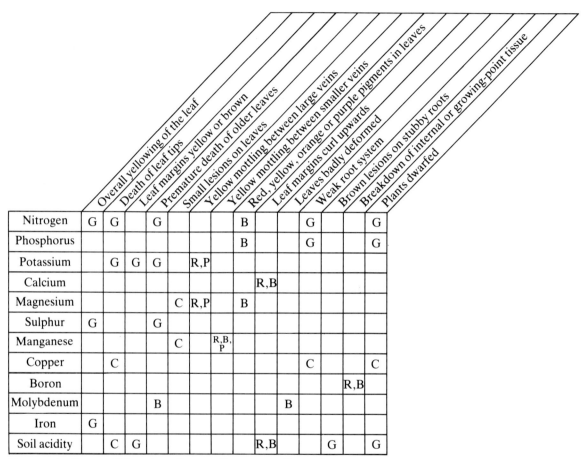

	Overall yellowing of the leaf	Death of leaf tips	Leaf margins yellow or brown	Premature death of older leaves	Small lesions on leaves	Yellow mottling between large veins	Yellow mottling between smaller veins	Red, yellow, orange or purple pigments in leaves	Leaf margins curl upwards	Leaves badly deformed	Weak root system	Brown lesions on stubby roots	Breakdown of internal or growing-point tissue	Plants dwarfed
Nitrogen	G	G		G		B					G			G
Phosphorus						B					G			G
Potassium		G	G	G	R,P									
Calcium								R,B						
Magnesium					C	R,P	B							
Sulphur	G			G										
Manganese					C		R,B, P							
Copper		C										C		C
Boron												R,B		
Molybdenum			B					B						
Iron	G													
Soil acidity		C	G					R,B			G			G

G-general, all crops; B-brassicas; R-roots; P-potatoes; C-cereals

Trace elements   The other main need for urgent treatment is associated with the occurrence of trace element deficiencies. A general guide to diagnosis is given in Fig. 15.8 but confirmation by a specialist adviser is essential. Some preventive and corrective measures against trace element deficiencies are given in Table 15.8.

Table: 15.8   *Preventive and corrective measures against trace element deficiencies*

Element	Preventive measures	Corrective measures			
		Chemical and rate of application	Method of application	Period of effective-ness	Precautions
Manganese	Avoid overliming	Manganese sulphate ($MnSO_43H_2O$) at 4–6 kg/ha	Spray on foliage immediately symptoms appear (or before if there is a known problem).	1 season	If applied in combination with herbicides check compatibility.
Boron	Avoid overliming	Borax ($Na_2B_4O_7$.$10H_2O$) 20–45 kg/ha	'Boronated' fertilizers applied to crops with high boron requirement.	1 season	Avoid application to crops susceptible to boron toxicity, especially cereals.
Copper	Avoid overliming	For soil application: copper sulphate ($CuSO_4.5H_2O$) at 20–50 kg/ha.	In solution to bare soil or in solid form to grassland.	5–8 years	Keep all stock off grassland until excess copper has been removed from herbage by rain.
Copper		For foliar application to arable crops: copper oxychloride at 3–5 kg/ha.	Spray on foliage as soon as symptoms appear.	1 season	
Molybdenum	Avoid soil acidity by normal liming programme.	Sodium molybdate (39% Mo) or ammonium molybdate (54% Mo) at 1.2–2.5 kg/ha.	In solution to the seedbed.	1 season	
Iron	Avoid overliming.	Iron-EDTA chelates 0.5–1.0 kg iron per hectare.	Spray on foliage as soon as symptoms appear.	1 season	
Zinc	Avoid overliming.	Zinc sulphate ($ZnSO_4.H_2O$) at 45–90 kg/ha.	In solution to bare soil.		
Cobalt	Avoid overliming.	Cobalt sulphate ($CoSO_4.7H_2O$) at 6 kg/ha.	In solution to pasture.	3–4 years	Consider cost/effectiveness compared with direct administration to stock.

Chapter 16 The economic use of fertilizers

Buying, storing and applying fertilizers accounts for some 10–50 per cent of the variable costs of producing a crop. This wide variation depends upon the effort needed to grow and harvest the crop successfully. For the hand-lifted potato crop, fertilizer costs are about 9 per cent of total variable costs. If the crop is machine-lifted the figure becomes 12 per cent. For cereal crops, on the other hand, fertilizer costs can account for 40–50 per cent of variable costs and any economies that can be made will appreciably increase profits.

There are many ways in which fertilizer costs can be reduced without adversely affecting yields or efficiency. Some may be adopted immediately with no change to the farming system. They include the effective selection and purchase of fertilizers, skilful interpretation of recommended rates of application and judicious timing and methods of application.

Other measures, such as the introduction of more leguminous crops into the rotation, may involve radical changes in the farming system although they can be adopted over a period of two or three years.

Finally there are important decisions which can be made in the knowledge that their benefits will develop slowly and affect the long-term fertility of the soil. They include all-important measures to stop the decline of organic matter in intensively cash-cropped soils and to build up the organic matter in others. Other possibilities include the use of cheap materials as storage fertilizers. This was done by previous generations using the then cheap and easily obtainable basic slag.

Immediate measures

The use of the cheapest effective fertilizers

It may be assumed that different water-soluble phosphorus and potassium sources, whether in compounds or in simple fertilizers, will have virtually the same effects on crop yield and quality. The same will be true for nitrogenous fertilizers on a broad range of non-calcareous soils provided that leaching is not a major problem (see Chapter 15, pages 202–203).

If the N, P and K in different fertilizers are assumed to be equivalent, decisions must be made on comparative costs. It is imperative that such comparisons should be made on the basis of unit prices (cost per kg N, P_2O_5 or K_2O) or by the other methods discussed in Chapter 15, pages 202–206. Comparisons made by using cost per tonne of fertilizer can be very misleading indeed. As a simple example, at the time of writing there is a 30 per cent difference in the unit cost (per kg N) of the dearest and cheapest simple nitrogenous fertilizers. On most soils the two would have the same effect, per kg N applied, on crop yields.

Simple versus compound fertilizers

Because of the costs of manufacture of compound fertilizers it is commonly cheaper to buy separate simple N, P and K fertilizers to supply a given amount of nutrients. As a rough guide, in terms of direct fertilizer costs, a grower spending £25 000 per year on compound fertilizer could save, at 1984 prices, some £2 200–£3 000 by applying the same amounts of N, P_2O_5 and K_2O as separate simple fertilizers. Against this must be set the very real conveniences of using a single compound to do the job, including uniformity of distribution and ease of application.

Nevertheless the use of simple fertilizers instead of compounds should always be considered. The economics must be worked out, not only in terms of direct costs of fertilizer, but also in terms of labour and machinery availability, rush periods of work and the cost and inconvenience of extra runs across the field.

The ability to use simple fertilizers blended together during spreading would greatly encourage their use by reducing three runs across the field to one. They cannot be applied together from a single hopper even if well mixed beforehand because of segregation of the three components during spreading. There are, however, already double-hopper spreaders on the market from which two simple fertilizers can be metered separately during distribution. It is certainly well within the capability of agricultural machinery manufacturers to produce a triple-hopper distributor to fulfil the requirements of accurate delivery of simple fertilizers to give an infinite variety of N : P : K ratios. This would ensure the 'tailoring' of closely specified fertilizer requirements and add to the savings because only one run over the field would be needed.

Types and time of application of nitrogenous fertilizers

Considerable savings in fertilizer costs may be made by reducing losses of fertilizer nitrogen by leaching or by volatilization of ammonia. The steps that may be taken are set out in detail in Chapter 15, pages 202–203, 207–209.

Interpretation of recommended fertilizer rates

Fertilizer recommendations are usually based on the assessed optima for crop yield. The main exceptions to this are where some major aspect of crop quality is concerned, as in malting barley. There are circumstances in which it is well worth while considering applications below the recommended rates, as follows:

Phosphorus and potassium
● If the recommendations under consideration are made by fertilizer merchants and exceed those made by an independent body such as ADAS.
● If soil P and/or K indices are high and the crop has low phosphorus or potassium requirements (e.g. P for sugar beet, P or K for cereals).
● If a policy of balance-sheet fertilizing is practised (see page 230) in soils with high P or K indices, some crops may be grown using only nitrogenous fertilizers provided the phosphorus and potassium removed by the crop are replaced at some point in the rotation.

Nitrogen More caution is needed in deciding whether to apply either less or more than the recommended rate of nitrogen for a given crop. The evidence of ill effects when exceeding the optimum on both yield and quality is now strong for all crops except those which give their yield as green leaves. There are, therefore, several advantages in not exceeding the recommended rates of nitrogen:
● Direct savings in fertilizer costs.
● Reduced leaching losses.
● Greater crop yield.
● Improved keeping quality and hence reduced storage losses.
    The safest procedure for most crops is to apply nitrogenous fertilizer at the rates recommended by ADAS, the Scottish

Agricultural Colleges or similar bodies after taking the greatest possible care to read the small print. In this connection ADAS have now published a computer programme, which, provided sufficiently accurate records are available, should greatly assist in deciding upon the N rates which will give the best economic returns.

The economic use of legumes

Easily the most neglected crops in the British Isles are the legumes, including potentially very profitable arable crops such as peas and beans, green manure crops such as lupins, forage crops such as lucerne and, above all, clover. There are many reasons for this neglect including management problems and moves towards over-simple farming systems such as continuous cereal culture. The prime example of the failure to use the nitrogen-harnessing power of legumes is found in the steady elimination of clover from swards for grazing or conservation. This has been done for the purpose of producing the last few kilograms of grass from each hectare of soil and, if this is to succeed, very high rates of nitrogenous fertilizers must be used. For those who can accept a lower level of production the use and maintenance of vigorous grass/clover swards will give major savings in expenditure on nitrogenous fertilizers, not only during the period of the ley but for arable crops over several years after it is ploughed. Even a one- or two-year ley rich in clover, when ploughed out, will give an N index of 2 for each of the next two years, implying a saving of 70–100 kg N/ha in each of those years as compared with the amounts of fertilizer required in continuous cereal cropping. Clover-rich 'permanent' pasture and leys of three or more years duration permit reductions in the rates of nitrogenous fertilizer for three, four or even five years after ploughing out.

Superimposed upon these savings are the very considerable reductions in nitrogenous fertilizer requirements for the grass/clover sward itself as compared with high N-input pure grass swards. These can be as much as 300 kg N/ha each year, but it must be firmly understood that the grass/clover sward will usually give a somewhat lower yield. The economics of changing over to a grass/clover system from high N-input pure grass swards should therefore be carefully considered by any farmer who is considering a less intensive, lower input approach to crop production.

If the farming system permits, peas and beans are excellent break crops with moderate requirements for fertilizer phosphorus and potassium and *none* for nitrogen. They are also regarded by ADAS as giving an N index of 1 for the next crop, thus reducing the fertilizer nitrogen requirements for a following cereal crop by some 40–50 kg N/ha as compared with that required by a cereal following another cereal crop.

Long-term action

Actions taken over a period of decades and built in to the farming system can greatly improve the general fertility of the soil and thereby reduce fertilizer costs while maintaining or improving crop yields.

Fertility build-up

The efficiency of fertilizer use, in terms of both reducing the amount needed by the crop and increasing the effectiveness of use by the plant, is overwhelmingly affected by:
● The cation-exchange capacity of the soil.
● The ability of the soil to hold water available to the plant.
● The reserve of nutrients in the soil.
All of these are strongly influenced by the type and content of clay and organic matter in the soil. The farmer can scarcely change soil clays but can improve the organic matter status considerably. The necessary measures are not always easy and are commonly neglected, resulting in a decrease in organic matter, sometimes with catastrophic effects such as wind and water erosion, surface capping, panning and increased leaching. All of these reduce fertilizer efficiency.

Any of the following will help to prevent a decline or bring about an increase in soil organic-matter content. Immediate results cannot be expected but long-term benefits over ten years or more will inevitably follow if the amount of humified soil organic matter is increased.
● Avoid continuous arable cropping, if possible.
● Introduce grass or grass/clover swards into the rotation. Obviously this is not always feasible but grass/clover leys if left for two years or more will contribute more to the organic matter content of cultivated soils than any other factors.
● If continuous arable cropping is unavoidable return every possible crop residue to the soil. This certainly includes cereal straw.
● Avoid excessive cultivations which assist the oxidative loss of organic matter.

● Maintain a green cover of the soil whenever possible.
● Grow and plough in green manure catch crops whenever this can be done without seriously impeding commercial cropping.
● Make and store farmyard manures efficiently and incorporate them into the soil immediately after spreading in order to avoid oxidative losses.
● Maintain an adequate soil pH by liming to encourage earthworm and micro-organism activity and produce good quality humus.

The use of storage fertilizers

Storage fertilizers are used to increase, as cheaply as possible, the reserve of a particular nutrient in the soil. They become available to the plant through the normal weathering processes.

The essential characteristics of storage fertilizers are:
● Cheapness.
● Low solubility in water.
● Slow availability to the plant over some years.
● Fineness of grinding to assist weathering processes.

The classical storage fertilizers are ground limestones (calcium and magnesium), basic slags (mainly phosphorus but also some calcium, magnesium and trace elements) and ground mineral phosphates (mainly phosphorus but also some calcium).

They may be used to maintain the reserves in soils already well supplied with a particular nutrient. A good example would be the use of ground mineral phosphate on soils of P index 2 or more simply to replace phosphorus taken off in crops, where little or no yield response would be expected to phosphate fertilizers. In such soil water-soluble fertilizers with only a very low phosphorus content may be used to meet any early-season needs of the crop for extra available phosphorus, supplemented by ground mineral phosphate applied at 0.4–0.8 t/ha once every four or five years for maintenance purposes.

In marginal or hill sites basic slags and ground mineral phosphates have been successfully used as the sole source of fertilizer phosphorus for pastures. For this purpose rates of application of basic slag vary from 0.6 to 1.2 t/ha once every 4–5 years, depending on the quality of the slag. Ground mineral phosphates should be applied at similar intervals at 0.4–0.8 t/ha.

For arable crops or intensive grass on soils with a low P index, basic slags or ground mineral phosphates should be used as supplements to water-soluble phosphates with the aim of gradually building up available phosphorus supplies.

Other potential storage fertilizers, not widely used, are gypsum (mainly sulphur but also some calcium); calcined magnesite (magnesium); crushed potassium-rich rocks containing slowly available potassium from micas or feldspars.

There are, at present, no nitrogenous storage fertilizers extensively used in agriculture.

Balance-sheet fertilizers

Because of inherent deficiencies of nitrogen, phosphorus and potassium and the consequent vigorous responses of crops to water-soluble fertilizers, at or around sowing time, a tradition has grown up of 'fertilizing for the crop'. This is based on sound principles for deficient soils as large supplies of immediately available nutrients are required.

Partly because of residues from former applications of fertilizers and manures there are many agricultural soils in which phosphorus and potassium reserves are high or very high (P and K indices of 2 or more). In these soils the yield responses of crops to phosphorus and potassium fertilizers are slight or non-existent and the need is simply to replace phosphorus and potassium taken off by crops to prevent a run-down in fertility. This presents an opportunity to fertilize, not for the individual crop, but for the rotation as a whole. Unfortunately this cannot be done with nitrogenous fertilizers because of their short-term effectiveness and the need for close annual control over available nitrogen.

For the other major elements, phosphorus, potassium, calcium, magnesium and sulphur, 'balance-sheet' fertilizing may improve fertilizer efficiency and cut costs without impairing yields. The approach can include the use of storage fertilizers described in the previous section. The aim should be to balance the nutrients removed in crops over a period of some years without doing so every year. Thus fertilizer nutrients other than nitrogen may be withheld from non-responsive crops and this can be compensated by applications to responsive crops without exceeding the estimated optimum rates. The cheapest sources of nutrients can often be used.

An example of a balance-sheet for fertilizers is given in Table 16.1. Its success depends upon detailed record-keeping and access to reliable information on the amounts of nutrients removed by crops. It must be stressed that the system does not apply to nitrogenous fertilizers and should not be used when dealing with soils deficient in phosphorus, potassium, sulphur and magnesium.

Table: 16.1   Examples of 'balance-sheet' fertilizing for two rotations.

Rotation 1*

Year	Crop	Yield (t/ha)	P_2O_5 removed in crop (kg/ha)	K_2O removed in crop (kg/ha)	Fertilizer P_2O_5 applied (kg/ha)	Fertilizer K_2O applied (kg/ha)
1	1 year lea (silage)	8 (dry matter)	40	120	Nil	Nil
2	Cereal	8	64	56	180	230
3	Cereal	8	64	56	Nil	Nil
4	Field beans	5	55	60	Nil	Nil
5	Cereal	8	64	56	180	180
6	Cereal	7	56	49	Nil	Nil
Total			343	397	360	410

Rotation 2*

Year	Crop	Yield (t/ha)	P_2O_5 removed	K_2O removed	Fertilizer P_2O_5 applied	Fertilizer K_2O applied
1	Cereal	9	90	108	Nil	Nil
2	Cereal	8	80	96	Nil	Nil
3	Oil-seed rape	8	120	80	250	360
4	Cereal	8	80	96	Nil	Nil
5	Cereal	7	70	84	Nil	Nil
6	Potatoes	40	40	240	250	360
Total			480	704	500	720

* In Rotation 1 it is assumed that all cereal straw has been ploughed in or burned.
  In Rotation 2 it is assumed that all cereal straw has been removed.

The most commonly used fertilizers in balance-sheet manuring are triple superphosphate (45 per cent P_2O_5) and potassium chloride (muriate of potash – 60 per cent K_2O). In some circumstances – if the soil pH is below 6.5 – it is feasible to use ground mineral phosphate (30–32 per cent water-insoluble P_2O_5) in place of triple superphosphate. The decision would depend upon soil type and comparative fertilizer costs. An adviser should be consulted.

The amounts of phosphorus and potassium removed by crops from soils with ADAS indices of 2 or more are rather greater than the average data quoted in Table 3.1 (p. 22). Approximate but sufficiently accurate data for the amounts of P_2O_5 and K_2O removed in 1 tonne of various crops at soil index 2 are given in Table

16.2. Balance-sheet fertilizing must always be somewhat rough and ready and errors of 20–30 per cent are not important so long as a periodic check on soil analysis is made – say every four years.

Table: 16.2   Approximate amounts of P_2O_5 and K_2O removed by crops growing in soils with ADAS indices of 2 or more.

Crop	Nutrients removed in 1 tonne of crop	
	P_2O_5	K_2O
Peas	10	10
Beans	11	12
Oil-seed rape	15	10
Sugar beet	2	8
Kale	8	22
Any cereal (grain)	8	7
Any cereal (straw)	2	5
Potatoes	1	6
Grass (dry matter)	5	15

Source: Data from several sources.

Allowance should be made for the phosphorus and potassium contents of any farmyard manure or slurry applied, using the *total* nutrient contents of the manures.

Balance-sheet fertilizers may be applied at any time in the autumn or winter preceding the crop for which they are immediately intended. The place in the rotation should be decided by considerations of crop response. For example it is essential in Rotation 2 to apply both phosphorus and potassium for the responsive potato crop.

Because of the ready uptake of potassium by the potato crop, there will be little residual potassium for the following two cereal crops. The decision not to apply further potassium for these two crops implies a deliberate run-down of the high soil potassium reserves until the next bulk application. This should not adversely affect yields.

Chapter 17   Fertilizers in the future

Amounts used   During the twentieth century, increases in crop production attributable to the use of fertilizers have been very great. In particular the use of much greater amounts of nitrogenous fertilizers, made practicable by advances in plant breeding, has given spectacular results. In fact responses of crops to fertilizers are greater, in cash terms, on many soils than those from almost any other source. It must not be assumed, however, that future increases in fertilizer use will necessarily result in increased crop production.

The amounts of NPK fertilizers being used annually in the world are increasingly vigorously. In areas such as the British Isles, with highly developed agriculture, the present increase is almost entirely in the form of nitrogenous fertilizers, the use of phosphorus and potassium fertilizers having ceased to increase some years ago. In less well developed areas nitrogen, phosphorus and potassium fertilizers are all increasingly used.

There is no sign that the increases on a world scale are slowing down and indeed there are still many areas where crop production could be dramatically increased by fertilizer use.

Even in Britain some crops, notably grass, receive less fertilizer than the optimum in many areas. There is no doubt, however, that the pressures over the last forty years of intensifying production towards maximum yields have led to excessive applications for some crops. This has led to inefficient and uneconomic use of fertilizer nutrients and has contributed to the pollution of watercourses by nitrates leached from the soil. It is most important also to appreciate that applications of fertilizer, particularly nitrogenous, in excess of the optimum requirement can depress yields from the maximum with consequent reductions in profit.

None the less such excesses will continue in intensive production systems so long as the goal of maximum yield is retained. Unfortunately, uncertainties about the true optimum rate of fertilizer, especially the lack of accurate methods of forecasting nitrogen requirements, encourage the use of excess fertilizer as an insurance measure, however falsely based this may be.

The question must now be raised as to whether the goal of maximum yield is achievable or indeed desirable. By controlling as many factors as possible we can produce crops which closely approach maximum yields but the input required is very great. In terms of the use of finite resources, such as the world's supply of phosphates and the fuels required to make nitrogenous fertilizers, the pursuit of maximum yield is not justified. This is extremely frustrating in the face of starving populations in many parts of the world but, without major changes, the world economic systems are not geared towards feeding people who are starving. In the present circumstances of surplus production in some parts of the world, grossly inadequate production in others and the economic problems of transferring food from one to the other, maximum crop production becomes futile.

Whatever the model of yield curve with increasing fertilizer rate for a given crop (Figures 13.3, 13.4, 13.5, Tables 13.2, 13.3, pp. 155–63, 157–159), the resources and energy expended on squeezing the last few kilograms of yield per hectare are being used very inefficiently and commonly uneconomically. From both the short-term profit motive and the long-term conservation motive, maximum yield production is inefficient.

Fertilizer efficiency

In terms of the proportions of fertilizer nutrients taken up by crops, the present range of fertilizers is inefficient. In general terms some 40–70 per cent of the nitrogen, 10–25 per cent of the phosphorus and 40–70 per cent of the potassium applied to crops in fertilizers are taken up by the first crop after application. In some circumstances recovery by the first crop can be greater than the top figures quoted here. One hundred per cent recovery of N and K by intensive grass and up to 50 per cent of P recovery by swedes have been recorded but such cases are unusual. Small amounts of residual phosphorus and potassium are taken up by subsequent crops over several years but most of the nitrogen not absorbed by the first crop is lost by leaching, denitrification or volatilization.

There is, therefore, scope for much improvement, but the problems are great, especially in improving the performance of nitrogenous and phosphate fertilizers.

Nitrogenous fertilizers

As already discussed, considerable improvement in nitrogen fertilizer efficiency could be achieved by cutting leaching and other losses, by careful timing of applications and by using formulations in which the easily leached nitrate is replaced by ammonium or urea nitrogen.

Slow-release and controlled-release nitrogen There are very sound reasons for developing a range of nitrogenous fertilizers with controllable and variable release rates. They might, for example, be used to avoid the need for several dressings of fertilizer nitrogen per year on intensive grass or on cereal crops. Annual crops such as potatoes could also benefit from a release of available nitrogen into the soil at or about the beginning of the grand period of growth. This would help to avoid excessive and inefficient foliage production early in the season which can happen all too easily if water-soluble nitrogenous fertilizers are applied at planting. A gradual release of nitrogen throughout the season would also prolong the efficient life of the foliage. It would be both necessary and feasible to combine some water-soluble material to assure early-season growth with controlled-release material to be activated later in the season.

There is a wide range of potential slow-release nitrogenous materials but most of them are, at present, produced on a very small scale and are much more expensive than traditional fertilizers. Some are used for high-value horticultural products such as container-grown plants where water-soluble fertilizers can easily cause salt and scorch damage. There is no doubt that they would be much cheaper if produced on a large scale. At present they are little used in agriculture because of their cost and problems about the rate at which they release available nitrogen.

As well as supplying available nitrogen at the time of maximum requirement, controlled slow-release nitrogenous fertilizers would most certainly reduce losses of nitrate by leaching and consequent stream pollution.

The present range of slow-release nitrogenous materials does not fulfil the requirements of controlled release within a few days of the time of need but does offer the possibility of reducing leaching losses. Slow-release materials fall into two main groups – synthetic organic materials and traditional fertilizers coated with other substances to prevent the immediate access of water.

The *synthetic organic materials* are nitrogenous substances resistant to breakdown and conversion to ammonium or nitrate by bacterial action. In this respect they are similar to much of the organic matter which reaches the soil as crop residues or manures.

Urea-formaldehyde resins are typical of this group. Their nitrogen content can be adjusted by varying the urea/formaldehyde ratio between 3/1 and 8/1, the highest ratio giving a product containing 35 per cent N. The rate of release of available nitrogen can also be adjusted to some extent by varying particle size – the larger the particles the slower the release. Thus, by using a range of particle sizes it is theoretically possible to ensure a steady rate of release of available nitrogen throughout a growing season. Although this may be achieved under controlled conditions in the greenhouse, variations of soil and weather make it much more uncertain in field crops. The rate of release in a particular soil, like the rate of mineralization of soil organic matter, is strongly influenced by soil pH, temperature and availability of water so that what might be slow release in one soil might be rapid in another. This is, at present, a major obstacle to the use of this group of materials to attain *controlled* release.

Other slow-release materials of this type, some of which have been used only experimentally are: iso-butylidene di-urea (IBDU) 32 per cent N; crotonylidene di-urea (CDU) 28 per cent N; oxamide ($NH_2CO.CONH_2$) 31 per cent N; and phthalamide ($C_6H_4(CONH_2)_2$) 17 per cent N.

All depend on particle size, soil pH and microbial activity for their rate of release. In fact they act in precisely the same way as hoof-and-horn, 14 per cent N, the natural organic slow-release nitrogenous fertilizer used for many years in horticulture.

Coated nitrogenous fertilizers at present available are mostly ordinary water-soluble fertilizers which have been coated with water-resistant bio-degradable materials. As such they suffer the same drawbacks as the synthetic organics in that the rate of release will vary from soil to soil and from season to season. The coatings may be resins which decay slowly. Sulphur has also been used to form a skin around urea particles and reduce the rate of nitrogen release. One such product consists of 3 parts of urea coated by one part of sulphur. It contains about 35 per cent N and 25 per cent S. In this case the rate of release will depend to some extent on the presence in the soil of a population of sulphur oxidizing bacteria. Such bacteria tend to thrive in acidic soils containing much humus.

Phosphorus fertilizers

The very low efficiency of phosphorus fertilizers is a direct result of fixation in the form of low-solubility phosphates of iron, aluminium and calcium. Very little phosphorus is leached from most mineral soils. The problem may be approached either by reducing the rate of fixation or by attempting to release available phosphates from the fixed state. Judicious liming reduces the rate of fixation in acidic soils. Also various formulations have been used to protect fertilizer phosphates from fixation. They have usually reduced the water solubility of the phosphate in some way – for example by the ammoniation of superphosphate – but have seldom been more effective than completely water-soluble materials.

Traditional water-soluble phosphates are derived from orthophosphoric acid (H_3PO_4). It is possible that they will be replaced to some extent by metaphosphates or polyphosphates derived from metaphosphoric acid (HPO_3) or more complex polyphosphoric acids. This group of substances are known as *condensed phosphates* and are potential fertilizer materials. Many of them are polymers, having large molecules with ring or chain structures. Because of this they need to be converted in the soil to orthophosphates before being taken up by the plant. There were hopes that this would slow down their activity as compared with ordinary water-soluble phosphates and that they would resist fixation. The present evidence is that this is not the case and that they are converted rapidly in the soil to orthophosphates – within hours or days – and thereafter act in the same way as traditional water-soluble phosphates. The use of condensed phosphates as fertilizers is, therefore, likely to depend more upon cost and phosphorus concentration than on resistance to fixation.

Surprisingly little attention has been paid to releasing the large amounts of residual phosphates, 75 to 90 per cent of the amounts applied each year. Until 1970, when phosphates suddenly increased in price, fixation could be accepted with equanimity as each crop could be treated with more cheap phosphates. As a result, there has been very little research on the release of fixed phosphates from the soil. It is possible that some lightly-fixed but still exchangeable phosphates could be released by the addition to the soil of powerful exchangeable anions such as fluoride (F^-), but much more radical treatment would be required to render soluble the more strongly fixed calcium, iron and aluminium phosphates.

One possibility is to use dilute mineral acids such as sulphuric acid, injected into the soil at points sufficiently far apart to minimize any toxic effects, which would dissipate rapidly and leave nuclei of temporarily water-soluble phosphates. Such treatment would also supply some sulphur to the soil and would release small pockets of available trace elements. Any such radical treatment would need very careful research before adoption and would also gradually deplete the soil phosphorus reserves but seems to be one of the few possible ways of rendering the reserves of phosphate available to the plant. Conservationists should remain calm about this proposal as the amounts of sulphuric acid used, for both practical and economic reasons, would be very small.

Controlled-release fertilizers

Coated fertilizers may be formulated to contain any nutrients, including the trace elements, and there is at least some hope that they may be widely used in the future. Apart from slowly degradable coatings other materials are being developed involving release controlled by osmotic pressure or by soil temperature.

We are, however, still a long way from producing true controlled-release fertilizers. The main interest of fertilizer manufacturers at present lies very much in the production and compounding of readily available water-soluble fertilizers and they have put vast amounts of capital into factories to do this. Yet the efficiency of these fertilizers is not high and there is much concern about leaching losses of nitrates. The use of controlled-release fertilizers is one of the very few aspects of agriculture in which research into fertilizer efficiency is likely to be rewarding in the next few decades.

The major problem is that of controlling release rates in widely different soils and seasons to give a reliable supply of a nutrient exactly when the plant needs it. It is a daunting task but by no means impossible.

The concentration of fertilizers

The concentration of nitrogen, phosphorus and potassium in fertilizers has increased almost threefold during the last forty years. Greater NPK concentration has reduced costs of transport and spreading appreciably, but has been achieved only by the virtual exclusion of magnesium and sulphur. Further concentration should be undertaken only if the advantages greatly outweigh the disadvantages.

The only real chance of increasing NPK concentration in solid fertilizers is by using the condensed phosphates described in the previous section. Some of them are also sufficiently soluble to be used to advantage in liquid fertilizers. Condensed phosphates have characteristically high P_2O_5 contents: calcium polyphosphate ($[Ca(PO_3)_2]_n$) has the equivalent of 64 per cent P_2O_5; potassium metaphosphate, which has the basic formula KPO_3, is a rich source of both P_2O_5 (60 per cent) and K_2O (30 per cent). These and other phosphates of calcium, sodium and potassium could obviously be used to increase the phosphorus concentration in fertilizers. There is little prospect of increasing the NPK concentration in solid fertilizers in any other way. At present advances in concentration are mainly confined to liquid or fluid fertilizers by the use of materials such as clays to maintain nutrients in suspension. As a result liquid fertilizers, previously much less concentrated than solids, are becoming more competitive in this respect.

Lime-magnesium-sulphur fertilizers

Elsewhere in the world, as well as in the British Isles, there is every sign that the major elements magnesium and sulphur are virtually neglected in fertilizers until deficiency conditions occur. This will no doubt continue and would probably be true also for calcium but for the accepted need for liming which ensures adequate supplies of this element.

The need for regular applications of magnesium is also now accepted in the British Isles but it is doubtful if the manufacturers, in their inexorable pursuit of NPK concentration, will re-introduce magnesium into the main range of compound fertilizers. It will, therefore, need to be applied separately. On acidic soils the cheapest form to use is magnesian limestone. On calcareous soils magnesium oxide (calcined magnesite) or magnesium sulphate (kieserite or Epsom salt) will be needed.

Despite many warnings and very strong evidence from parts of the British Isles and other areas of the world where atmospheric sulphur pollution is slight, the need for regular applications of sulphur in fertilizers is not yet widely accepted in this country. It certainly will be within the next decade and the case will become more urgent if anti-pollution measures become effective.

The fertilizer manufacturers can meet some of the sulphur requirements of crops by using potassium sulphate and perhaps some ammonium sulphate in compound fertilizers but there will be some crops and soils for which that would not be enough and there will be a need to use cheap sources of sulphur such as gypsum ($CaSO_4.2H_2O$).

Unless the manufacturers of traditional NPK fertilizers can move to meet the sulphur and magnesium requirements there is a very strong case for a new range of fertilizers, complementary to the NPK compounds, to supply these neglected major elements. At the same time, because of more intensive farming, higher crop yields and the use of acidifying fertilizers, there is also an increasing need to consider annual or biennial applications of lime.

In these circumstances a new range of fertilizers may be developed to supply lime, magnesium and sulphur. Unlike NPK fertilizers the materials used would be much less soluble in water and would be in the nature of storage fertilizers. The lime, magnesium and sulphur could be varied according to need, using cheap materials such as calciferous limestones, chalk, magnesian limestones, calcined magnesite and gypsum. Such a new range of non-NPK fertilizers could replace liming as we know it and would also supply the neglected major elements, sulphur and magnesium. The aim would be to apply some 500–2 000 kg/ha annually or biennially according to needs to complement NPK fertilizers. The relatively slow availability of the calcium, magnesium and sulphur would not be a drawback so long as regular applications were made. Applications could be made using normal farm equipment and, as timing would not be critical, they could be made at any convenient time.

Trace elements in fertilizers

Most soil/plant workers have consistently opposed 'blunderbuss' treatments with trace elements, either as sprays or included in fertilizers, for the following reasons:

- The relatively rare incidence of trace element deficiencies during the period 1900–1960.
- Rapid fixation of soil-applied trace elements, particularly manganese, and the resultant lack of yield response in crops.
- Problems of uneven distribution, in the field, of the small amounts of trace elements required.
- Scorching of foliage by multiple-element sprays.
- The narrow margins between deficiency and toxicity levels, especially of boron.

Increased crop yields together with reductions in trace element availability resulting from liming have now led to widespread deficiencies of some trace elements. There is strong evidence that deficiency of manganese is now endemic in many soils. Deficiencies of boron and copper are increasing and will continue to do so. There are also strong theoretical grounds, based on the total zinc content of soils and crops, to forecast that zinc deficiency, so far negligible in the British Isles, will occur on certain soils within the next decade. The most likely areas are those with soils derived from limestones, sandstones or granite.

The present practice of spraying crops with solutions containing the appropriate trace elements, either as a precautionary measure or on the appearance of deficiency symptoms in the crop, can give spectacular results and will undoubtedly continue. There is, however, at least a reasonable case for including trace elements in compound fertilizers provided that the problems of rapid fixation in the soil, as well as machinery corrosion in factories and distributors, can be overcome. Care would also be needed, in research and formulation, to avoid toxic levels, particularly of boron for cereal crops where the margin between deficiency and toxicity is so slight.

One possible approach would be to incorporate manganese, boron, copper and zinc in granular compound fertilizers. The use of finely ground but sparingly soluble forms of the trace elements, such as the oxides of the three metals and less soluble borates, would help to overcome the risk of trace element toxicities. The acidifying nature of most modern fertilizers would help to counteract the slow release of the trace elements in these forms by producing a small acidic zone around each fertilizer granule in the soil persistent enough to dissolve and render available the trace element for some weeks. Thus, the acidity of the fertilizer would help to overcome the rapid fixation which occurs when trace elements, especially manganese, are applied directly to the soil in either powders or sprays.

The granular nature of compound fertilizers would seem to be essential to the creation of these small acidic zones and this could not occur if liquid fertilizers were used.

Other possible approaches to the fixation problem would be through the development of coated controlled-release fertilizers or by the production of new formulations of fritted trace elements in glass-like substances. Whatever approach is finally adopted it is likely that compound fertilizers containing a range of trace elements will be much more widely used within the next ten years.

The use of simple fertilizers

The use of simple fertilizers, prevalent before 1940, has been partly eclipsed by the use of compound fertilizers. The only major exception is that of simple nitrogenous fertilizers based mainly on ammonium nitrate or urea.

The present position of compound fertilizers in the British market is apparently unassailable – so much so that the only simple fertilizers listed by some major manufacturers are nitrogenous.

Yet there are advantages in both price and formulation in using simple fertilizers. Each manufacturer chooses to produce a limited number of compounds and, although competition has obliged them to produce a range with several nutrient ratios, the full spectrum of N : P_2O_5 : K_2O ratios can be produced in practice only by blending simple fertilizers. Bulk blending and storing is difficult in the British Isles because of the high humidity and consequent caking of the blended fertilizers – the cementing of individual granules into lumps.

Blending of simple fertilizers during distribution in the field is now distinctly feasible. Double-hopper machines are already on the market and it would be a short step from them to produce a triple-hopper spreader with individual metering devices for each hopper. Using such machines simple nitrogen (urea, ammonium nitrate), phosphorus (e.g. triple superphosphate) and potassium (potassium chloride, potassium sulphate) fertilizers could be blended in an infinite range of ratios during distribution, thus giving precise formulation and saving the extra costs usually in-curred in the use of compound fertilizers. If this came about and trace element inclusion were needed, as suggested in the previous section, these elements could be included by the manufacturer in the nitrogenous components.

It is ironic that the final paragraphs of this book should recom-mend a return to practices abandoned forty years ago but present technology and economics both in fertilizer manufacture and fertilizer-spreading machinery point strongly towards such a move.

Conversion table (metric – imperial)

Weight	1 kilogram (kg)	= 2.2 lb
	50 kilograms (kg)	= 110.25 lb
	1 tonne (t)	= 2205 lb or 0. 984 tons
Length	1 metre (m)	= 1.09 yards
	1 kilometre (km)	= 1094 yards or 0.62 miles
Volume	1 litre (1)	= 1.76 pints or 0.22 gallons
Area	1 hectare (ha)	= 2.47 acres
Weight/Area	1 kg/ha	= 0.9 lb/acre
	50 kg/ha	= 0.4 cwt/acre or 1 cwt/ha
	1 tonne/ha	= 0.4 tons/acre
Volume/Area	1 litre/ha	= 0.09 gallons/acre

Further reading

Russell, E W 1980 *Soil Conditions and Plant Growth* 12th edn. Longman

Soils and plant nutrition

Simpson, K. *Soil* Longman

Fertilizers

Cooke, G W 1982 *Fertilizing for Maximum Yield* 3rd edn. Granada

Finck, Arnold 1982 *Fertilizers and Fertilization* Verlag Chemie

Fertilizer recommendations

ADAS* (Ministry of Agriculture, Fisheries and Food) 1984 Booklet 2191 *Lime and Fertilizer Recommendations No. 1 Arable Crops*

ADAS (Ministry of Agriculture, Fisheries and Food) 1984 Booklet 2430 *Lime and Fertilizer Recommendations No. 5 Grass and Forage Crops*

East of Scotland College of Agriculture* 1983 Bulletin No. 28 *Fertilizer Recommendations* 3rd edn.

* These organizations also publish numerous booklets or leaflets of a very high standard dealing with many specific aspects of crop production, fertilizers, lime, manures and soil.

Index of definitions and descriptions

General Index

acid rain, 56–7
ADAS
  fertilizer
    recommendations, 165–6,
    186, 187, 188–93, 230–2
adverse effects of fertilizers, 8–9,
  77–8, 79–80, 126, 127–9, 145–
  7, 150–1, 174–5, 179–84, 217–
  18
adverse effects of lime, 37–8, 54–6,
  71–3
aluminium toxicity, 70
ammonia
  aqueous, 130
  in liquid fertilizers, 115
  injection into soil, 128–9, 147,
  220
  loss by volatilization, 91, 92,
  126, 130
  soil acidification by, 129
annual losses of nutrients, 9, 79
antagonism, 32, 37–8, 162, 176
atmosphere, 56–7, 167
availability of nutrients, 28–43,
  46–64
  and cation exchange, 34–6
  and green manures, 105–8
  and liming, 54–6
  and rock weathering, 34
  and root ramification, 49–50
  and soil acidity, 36–7

assessment, 59–64
concept, 28–31
cycle, 29
in manures, 83–4, 89–91, 93–5
positional, 30–1
prediction, 50–1
restricting factors, 31–4
variability, 29

'balance sheet' fertilizers, 226, 230–
2
Big bags, 207, 212–13
biological oxygen demand
  (B.O.D.), 9, 87
blended fertilizers, 110, 113, 225–6,
  242
boron, 20
  deficiency, 71–3, 180
  in sewage sludge, 100
brassica crops
  sulphur requirement, 19
  symptoms of acidity, 70

calcareous soils
  and fixation of trace elements, 42
  and phosphorus fixation, 49
  pH, 65
calciferous limestones, 53, 73–4, 138
calcium, 19–20, 38

availability and leaching, 53–4,
  77–9
deficiency, 19–20
in fertilizers, 112, 138–9
in limestone, 53, 73, 138–9
in manures, 90
reserves, 53–4
role in nutrition, 19–20
calculations
  comparative fertilizer
    prices, 204–6
  fertilizer requirements, 197–202
  rates of application, 197–202
  unit price, 204–5
cation exchange, 34
  and nutrient availability, 32,
  34–6
cation exchange capacity, 34–5
  and lime requirement, 67–8
cereal crops
  continuous, 170
  fertilizer nitrogen for, 162,
  177–9, 207–9
  quality, 177–9
  trace element deficiencies, 71–2
  yield response to farmyard
    manure, 88
chalk, 4, 52, 53, 66–7, 73–4, 138
clays
  cation exchange capacity, 34–5
  phosphorus fixation by, 32, 42